D.H.

CONVERGENT EVOLUTION IN CHILE AND CALIFORNIA

US/IBP SYNTHESIS SERIES

This volume is a contribution to the International Biological Program. The United States' effort was sponsored by the National Academy of Sciences through the National Committee for the IBP. The lead federal agency in providing support for IBP has been the National Science Foundation.

Views expressed in this volume do not necessarily represent those of the National Academy of Sciences or of the National Science Foundation.

US/IBP SYNTHESIS SERIES 5

CONVERGENT EVOLUTION IN CHILE AND CALIFORNIA

Mediterranean Climate Ecosystems

Edited by

Harold A. Mooney
Stanford University

Dowden, Hutchin∫on & Ross, Inc.
Stroudsburg   Pennsylvania

LIBRARY OF CONGRESS CATALOGING IN PUBLICATION DATA
Main entry under title:
Convergent evolution in Chile and California: mediterranean climate ecosystems
    (US/IBP synthesis series; 5)
    Bibliography: p. 000
    Includes index.
    1. Convergence (Biology) 2. Ecology—California. 3. Ecology—Chile. I.
Mooney, Harold A. II. Series.
QH373.C66   575   77–1884
ISBN 0–87933–279–4

Exclusive distributor: **Halsted Press**
A Division of John Wiley & Sons, Inc.
ISBN: 0–470–99227–1

FOREWORD

This book is one of a series of volumes reporting results of research by U.S. scientists participating in the International Biological Program (IBP). As one of the 58 nations taking part in the IBP during the period July 1967 to June 1974, the United States organized a number of large, multidisciplinary studies pertinent to the central IBP theme of "the biological basis of productivity and human welfare."

These multidisciplinary studies (Integrated Research Programs), directed toward an understanding of the structure and function of major ecological or human systems, have been a distinctive feature of the U.S. participation in the IBP. Many of the detailed investigations that represent individual contributions to the overall objectives of each Integrated Research Program have been published in the journal literature. The main purpose of this series of books is to accomplish a synthesis of the many contributions for each principal program and thus answer the larger questions pertinent to the structure and function of the major systems that have been studied.

Publications Committee: US/IBP
Gabriel Lasker
Robert B. Platt
Frederick E. Smith
W. Frank Blair, Chairman

PREFACE

This book reports the results of a study on the convergent evolution of mediterranean-climate ecosystems, supported by the International Biological Programs of the United States and Chile. This study was conducted over a several year period and was a truly international venture involving the participation of scientists from many universities in the United States as well as Chile.

The results of this project are given in three parts. The first part is a review of the existing knowledge on mediterranean-climate ecosystems, which served as a model for our study of convergent evolution. Scientists from most of the mediterranean-climate regions of the world reported on specialties ranging from geography through biogeography and ecology at a conference in Valdivia, Chile. The proceedings of this conference were published in 1973 (di Castri and Mooney).

The new knowledge gained during this project is given in two additional volumes. One is an *Atlas* (Thrower and Bradbury, 1977) and the other the present volume which is a *Synthesis*. These two volumes are designed as a single integral unit with minimal overlap. The *Atlas* gives much of the basic data gained during the study in a highly organized format but with little interpretation.

The *Synthesis*, which is based in large measure on the detailed information furnished by the *Atlas*, represents our view of the significance of these findings based on the state of current knowledge. The existence of the *Atlas* ensures that reinterpretations can be undertaken when further information becomes available, from a data base uncompromised by any of the subjectivity or speculation that necessarily finds its way into this *Synthesis*.

As is outlined in the introductory chapters, this *Synthesis* (and the companion *Atlas*) bring together wholly new information, from precisely defined, closely matched, and intensively studied areas, in an attempt to bring some resolution to the central ecological question, "Are there limited evolutionary solutions to a given environmental complex?"

The names of many of the scientists and technicians who participated in this program are given in the Appendix. We all extend our appreciation to the U.S. National Science Foundation for generous support of the program from start to finish and to W. Frank Blair and Francesco di Castri, who were instrumental in the early planning phases. William Carver aided considerably in the editorial process for the *Synthesis* and Peter Raven and Robert Whittaker in its review. Noël Lallana Diaz prepared the illustrations on chapter-opening pages.

Finally, Otto T. Solbrig and I express our appreciation to the Guggenheim Foundation for support during the synthetic phase of the project.

Harold A. Mooney
Stanford, California

CONTENTS

LIST OF
CONTRIBUTORS

Homer Aschmann
Department of Earth Sciences, University of California, Riverside

Conrad Bahre
Department of Geography, University of California, Davis

David E. Bradbury
Department of Geography, University of Arizona

Celia Chu
Department of Biological Sciences, Stanford University

Martin L. Cody
Department of Biology, University of California, Los Angeles

Eduardo R. Fuentes
Laboratorio de Ecologia, Universidad Catolica de Chile

Juan Giliberto
Laboratorio de Botanica, Universidad Catolica de Chile

William Glanz
Department of Zoology, University of California, Berkeley

Ernesto Hajek
Laboratorio de Ecologia, Universidad Catolica de Chile

Rachel I. Hays
Department of Biology, San Diego State University

Adriana Hoffmann
Laboratorio de Botanica, Universidad Catolica de Chile

James H. Hunt
Department of Biology, University of Missouri

Albert W. Johnson
Department of Biology, San Diego State University

Sterling Keeley
Department of Botany, University of Georgia

Jochen Kummerow
Department of Botany, San Diego State University

Valmore LaMarche
Tree Ring Laboratory, University of Arizona

Philip C. Miller
Department of Biology, San Diego State University

Andrew R. Moldenke
Board of Studies in Biology, University of California, Santa Cruz

Harold A. Mooney
Department of Biological Sciences, Stanford University

David J. Parsons
National Park Service, Sequoia and Kings Canyon National Parks

Otto T. Solbrig
Department of Biology, Harvard University

Norman J. W. Thrower
Department of Geography, University of California, Los Angeles

CONVERGENT
EVOLUTION
IN CHILE AND
CALIFORNIA

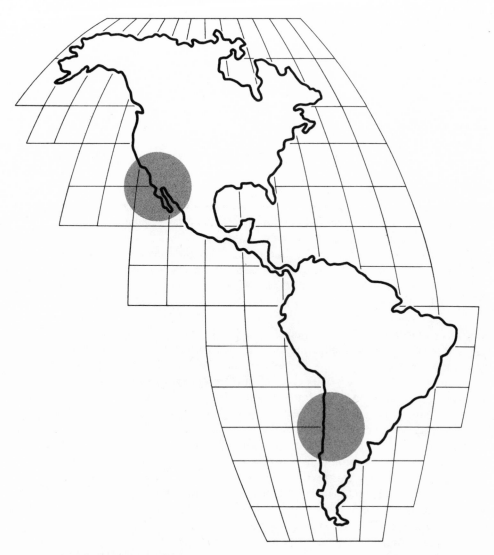

Introduction

H. A. Mooney
O. T. Solbrig
M. L. Cody

In recent years there has been a renewed awareness of the importance of ecosystem studies. One result of this reawakening is the decision by the U.S. sector of the International Biological Program (IBP) to direct a major part of its effort toward a number of such studies. The California/Chile project synthesized in this volume is but one of these: it is a study of whole ecosystems, but further it attempts to balance ecological conclusions with evolutionary interpretation.

LIMITATIONS OF THE ENERGY-BUDGET APPROACH

Ecosystem studies proceed in various ways. The approach that has gained broadest acceptance involves the development of energy budgets for a particular community, or ecosystem. Primary productivity, standing biomass, and the pathways through which energy flows are all measured very accurately. With this information in hand, as well as data on the independent variables of the system—primarily temperature, rainfall, and insolation—the ecologist produces a numerical (computer) model of the ecosystem. Using the annual cycle of independent variables unique to that ecosystem, the model then predicts primary and secondary productivity.

There are very good reasons for adopting such an approach to ecosystem studies. Energy is essential for the operation of ecosystems, and the patterns and cycles of energy flow are basic attributes of any particular system. Furthermore, the models thus evolved should prove to be of great value in the development of management techniques directed toward the maximization of food production for human consumption without concomitant habitat degradation.

The basic limitation of the approach outlined is that there is no theoretical or experimental reason to assume that a model evolved for one system or area is generalizable to another. In undertaking a number of parallel-biome studies involving different vegetational types, the U.S. IBP implicitly recognized this limitation. We have no a priori reason to expect that a model of a grassland developed for, say, the short-grass prairie of Colorado will have relevance in, say, the Argentine pampas. The problem lies in the nature of the usual energetic model: the properties of the region's component species are taken as given, and the immediate consequences of these properties are then investigated. But because the properties are in fact not static, but evolving, the system will respond to management techniques not only by altering its patterns of energy flow but also by forcing changes in the basic properties of its species. Consequently, we need alternative approaches to ecosystem studies that will allow us to predict the changes in the structural and functional properties of the component species *that can be expected* under conditions of land management or despoliation.

AN EVOLUTIONARY APPROACH
TO ECOSYSTEM STUDIES

The fundamental question asked in our study is whether two very similar physical environments in different parts of the world, acting on phylogenetically dissimilar organisms, will produce structurally and functionally similar ecosystems. If we cannot answer in the affirmative, the predictive power of community ecology will be weakened. In effect, knowledge acquired from studying one particular ecosystem will not be relevant to another unless similar physical environment indeed means similar ecosystem.

It has been known for a long time that in areas of climatic similarity, however separated the areas are geographically, the physiognomy of the vegetation is also similar. These similarities have encouraged the classification of vegetation into broad general types, the types coinciding more or less with certain climatic attributes (Schimper, 1898; Du Rietz, 1931; Raunkier, 1934; Cain, 1944; Good, 1956). Intuitive satisfaction with these schemes has led in turn to attempts to predict vegetation using solely climatic features (Köppen and Geiger, 1930; Holdridge, 1947; Tosi, 1960), with broadly predictive results. Although the details of the different approaches are subject to question, the premise that climate is a major determinant of vegetational physiognomy is generally accepted. From this premise follows the corollary that environmental similarity will produce similarity in vegetational structure independent of the lineages supplying the floras in question.

If similar physical environments do indeed produce similar ecosystems, then each combination of physical and climatic factors (temperature, humidity, insolation, exposure, soil type, etc.) evidently produces a unique set of physical stresses, to which there exists only a limited number of successful adaptive strategies. Otherwise, different solutions would obtain: unlike ecosystems would emerge in like circumstances. What is called for, then, is the identification of the various stresses created by the environment and the prediction of those adaptive strategies that will accommodate not only these stresses but also any new stresses the strategies themselves may create.

This approach, namely, the identification of the adaptive strategies of the various members of the ecosystem and of the selective forces that account for these strategies, is what we choose to call the evolutionary approach to ecosystem study. It implies that for each set of environmental conditions there is an energetic benefit and a cost to the various structures and functions that an organism can adopt, and that natural selection will choose those structures and functions that maximize Darwinian fitness. The approach thus leads to the articulation of a theory explaining why similar physical environments should indeed lead to similarly structured ecosystems.

THE RESEARCH DESIGN

The basic hypothesis, that areas of the world with similar climates will support communities with comparable structural features, even where the two sets of component organisms have had diverse evolutionary histories, was tested by choosing two areas of the world that are distantly separated and yet have similar climates and a close match in physiography, geology, and land-use features. One of these areas was in southern California, the other in central Chile. Both have a mediterranean-type winter-rain summer-drought climate. In each of these regions a central station was selected for intensive investigation; each of these two stations, the *primary sites*, is centered within the region of true mediterranean climatic type. Additional stations, the *secondary sites*, nearby but representing moister or drier variants of the cimate of the primary site, were also selected in both regions. Thus the design allows us to test the basic hypothesis not only in the areas of homologous but disjunct climate but also in adjacent areas of dissimilar climate.

The primary site chosen for Chile (see Figure 1-1) was Fundo Santa Laura, located approximately 30 km northwest of Santiago; for California, we chose Echo Valley, near San Diego (see Figure 1-2). Sites representing the nearby montane climates were located at Cerro Roble, Chile, and Mount Laguna, California (Table 1-1). For coastal climatic-research stations we selected the Papudo/Zapallar/Cachagua region of Chile and Camp Pendleton, California. Finally, xeric climatic sites were established at Cerro Potrerillo in Chile and Cabo Colnett in Baja California.

The primary sites were used for our intensive quantitative comparisons of environment and community structure. Here we studied in detail not only the past and present climates, the geology and soils, the land-use histories, but also the community structure and the temporal and spatial response of organisms to a well-defined resource network.

The secondary sites were utilized for the necessary control comparisons of community structure between adjacent areas of climatic dissimilarity but phylogenetic similarity. We could not engage in studies at the secondary sites of the full range of detail as those at the primary sites, but their macroenvironments and community structures were characterized. There was some small shifting of secondary sites during the course of the program for various reasons. Torrey Pines State Park, which was originally designated as the California coastal secondary site, was replaced by Camp Pendleton Marine Corp Base to facilitate ease of operation. The coastal "site" in Chile had to be designated as a local region (Papudo/Zapallar/Cachagua) because of the patchy nature of the natural vegetation. Also here the project climatic station was moved because of a fire during the course of the study.

Additional sites (Table 1-1, Figures 1-1 and 1-2) were selected for specialized aspects of the program as discussed in later chapters.

Although the study sites were selected on the basis of degree of similarity

FIGURE 1-1. *Study region of central Chile.*

of the physical environments, our options were limited to areas for which complete and reliable long-term records of the macroclimate were available. The selection of specific sites within these areas was based on qualitative observations of similarity in topography, geology, and land use. It was vital, however, that we determine precisely the degree of similarity of the matched primary and secondary sites. Accordingly, in-depth geographical and climatic

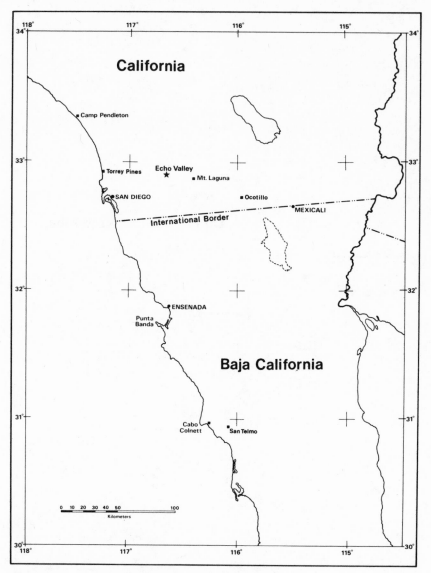

FIGURE 1-2. *Study region of southern California.*

studies of landforms and land use and of macro- and microclimates have been an integral part of the project.

That the climatically matched primary and secondary stations had at least comparable vegetation physiognomies was immediately obvious (Figure 1-3). The vegetation encountered along our climatic gradient ranged from forests through various scrub community types (Table 1-1), and they appeared closely matched intercontinentally at equivalent climates.

TABLE 1-1 *Study Localities*

Station	Geographic position		Elevation, m	Estimated annual precipitation, mm	Dominant vegetation
	Longitude	Latitude			
California					
Primary site					
Echo Valley	116°39'W	32°54'N	1070	550	Sclerophyll scrub (chaparral)/ evergreen woodland
Secondary sites					
Mount Laguna	116°26'W	32°50'N	2030	500	Montane forest
Camp Pendleton Marine Corps Base	117°31'W	33°21'N	90	225	Coastal scrub
Cabo Colnett	116°03'W	30°45'N	40	140	Coastal succulent scrub
Additional sites					
San Telmo	116°06'W	30°58'N	175	160	Coastal succulent scrub
Punta Banda	116°44'W	31°49'N	100	250	Coastal succulent scrub
Ocotillo	116°00'W	32°44'N	130	100	Desert scrub
Torrey Pines State Park	117°15'W	32°55'N	110	200	Coastal scrub/sclerophyll scrub/ coastal pine
Chile					
Primary site					
Fundo Santa Laura	71°00'W	33°04'S	1000	600	Sclerophyll scrub (matorral)/ evergreen woodland
Secondary sites					
Cerro Roble	71°30'W	32°58'S	2222	750	Montane forest
Papudo/Zapallar/Cachagua	71°28'W	32°35'S	100	350	Coastal scrub
Cerro Potrerillo	70°43'W	30°20'S	100	125	Coastal succulent scrub
Additional sites					
Los Molles	71°32'W	32°13'S	20	300	Coastal succulent scrub
El Tofo	71°30'W	29°29'S		100	Desert scrub

FIGURE 1-3. *Aspect of the vegetation of matched climatic stations in California and Chile. Top—montane climate; left, Cerro Roble, Chile; right, Mount Laguna, California; center—interior mediterranean climate; left, Fundo Santa Laura, the Chilean primary site; right, Echo Valley, the Californian primary site; bottom—coastal xeric mediterranean climate; left, Cerro Potrerillo, Chile, Cabo Colnett, Baja California.*

Within this research design the principal thrust was the study of the structural attributes and certain functional attributes of selected members of the two homologous communities. In selecting groups of organisms or features of the ecosystem for study, we sought to encompass a wide range of ecosystem phenomena. It was thus necessary to sample various components within each ecosystem rather than to undertake its total analysis. For although it might be *desirable* to learn whether these disjunct ecosystems are comparable in all details of structure and function, to do so would be essentially impossible within our constraints of time and resources; whereas it would be *sufficient* (and possible) for our overall objectives to conduct intensive studies on a carefully selected range of topics.

The problems of analyzing similarities and differences between species are formidable and differ between various taxonomic groups. In characterizing an organism one can describe its anatomy, morphology, life history, physiological responses, distribution patterns, and so forth. For plants we have made comparisons in most of these categories for a large number of species. Thus our assessment of the degree of convergence in plants is based on the similarity of a wide range of features, and the yardstick of comparison is the match with characteristics of plants inhabiting dissimilar environments.

In animals in the field many of the readily noticeable and more memorable traits such as method of locomotion, color pattern, behavior, and vocalization are traits that exhibit an astounding variability among animal species of all sorts, but that scarcely exist as variables in plants, which are routinely green, stationary, and lacking in behavior. Hence, with such animal traits as these, all of them plastic and influenced by a variety of selective forces, we confront many possibilities of nonconvergence or merely approximate convergence.

In addition, studies on animals are less amenable to experimental techniques than those on plants. We can amass, for example, the life-history details of reproductive effort or the economics of changing physiological performance with changing ambient environment much less easily in the animals we studied than in plant species. It is therefore appropriate to summarize the special problems, and to indicate their solutions, in our tests for convergent evolution in consumer organisms. These problems are fundamental to studies of this sort; they need to be aired if our conclusions are to find much acceptance; and they will undoubtedly recur as more studies of this nature are conducted.

Some of the difficulties with the tests are those of much of nonexperimental science: theories are generated, substantiated, and eventually established through repeated examination of certain facts, regularities and measurements that corroborate what is expected if one provisionally adopts the working hypothesis. In the present case there are two sorts of difficulties, one trivial and the other not so trivial.

The trivial difficulty lies in the fact that the studies are nonexperimental. An experimental test of convergent evolution might, for example, be thought to require the introduction of a novel, exotic family of lizards into a habitat set of some species-depauperate mediterranean region. But since nature has provided exactly such "introductions," using largely different taxa with different speciation possibilities on different continents, we need not duplicate these efforts with our own. And, in any case, the practical problems of such an experiment are overwhelming. Moreover, man has brought about a good many inadvertent introductions—some of them centuries ago—of cosmopolitan or tramp species (far fewer animals than plants among them, however), and their fate in communities rich or depauperate in their closet competitors

bears out conditions and expectations. Conceivably, small-scale experimenta-
tion could be conducted (putative ecological analogs could be pitted against
one another, so to speak, in enclosures of some sort), but, again, the returns
seem scarcely likely to warrant the effort.

A more formidable potential difficulty lies in labeling two phenotypes
or phenotypic traits—behaviors, colors, morphologies—similar and two others
dissimilar. How similar is similar, and what level of similarity constitutes
evidence for convergent evolution? This difficulty is like that of classifying
much of the world around us, for instance the naming of colors: their names
are learned by contrasts, and they attain meaning only by virtue of the relative differences in reflectance of surfaces. This problem is particularly acute
for those who learn facts secondhand; but for the field worker with a wide
repertoire of experiences it does not exist, for he immediately perceives and
puts into perspective the similarities and differences, the achievements and
failings, of selection for convergence.

The difficulty is largely overcome by increasing the variety and number of
the contrasts made. A study concluding that unrelated species A and B are
ecologically and morphologically similar can make its point in several ways,
by introducing other contrasts in just the same way that cardinal red and
burgundy red have a similarity emphasized by relating these colors to, for
instance, apple green. Thus A can be contrasted to close genetic relatives
A' that occupy nonmediterranean habitats (or different niches in mediterranean habitats) on the same continent, and B likewise, and the level of
similarity between A and B can be contrasted to that between A-A' and
B-B'. Or A and B could be compared in some way to the chances of selecting
at random a pair of species from the pool that contains a range of all species,
including those as distantly related as are A and B. Or, finally, they might be
compared to all species in mediterranean habitats with the same degree of
genetic relationship.

We can illustrate these alternatives using the pigeon guilds, birds in the
family Columbidae that occur in mediterranean habitats on all five continents as well as in many other sorts of regions. When we census mediterranean
habitats in California, we find just two species: the small *Zenaidura macroura*
(wing, 150 mm), of a long-tailed character and plumbaceous hue, which
occurs most commonly in lower and open habitats; and a larger species,
Columba fasciata (wing, 208 mm), grayish in color, which occurs in woodland
and forest. We choose to accept the hypothesis of convergent evolution here,
because in Chile we find a long-tailed plumbaceous pigeon (*Zenaidura auri-
culata*; wing, 151 mm) in lower and more open habitats, and a larger, grayer
bird (*Columba araucana*; wing, 211 mm) in forests, and no other pigeons on
the mediterranean habitat gradient. Naturally, there are differences between
the members of these two pairs of species analogs, as well as similarities; both
the similarities and the differences help the taxonomist to classify the species

as congeneric or not, and as allospecific. But what is the significance, if any, of the remaining differences of 1 or 2 mm in wing lengths? How can we objectively assess this difference? We could list the Californian pigeons, seven species, ranging from 87 to 230 mm in wing lengths, and say something about the chances of selecting our two at random from this set. We could likewise cite the range in wing lengths of the eight Chilean pigeons as 91 to 211 mm, and guess the chances of getting two pairs of cross-continental matches that included our four birds. But all of this seems a rather pedantic exercise, which pays lip service to the scientific method while advancing its ends very little.

To emphasize the similarity in the two pairs of mediterranean-zone birds mentioned, we can cite a third data set. In the mediterranean habitats of southern Sardinia one also finds two columbid species (Walter and Cody, unpubl.): one is small (*Streptopelia turtur*; wing, 175 mm), plumbaceous but not conspicuously long-tailed, and occurs in lower, more open habitats; the other is larger and gray (*Columba palumbus*; wing, 240 mm) and occurs in the taller woodlands! Again, we could mention the seven other European species, and quote odds on finding just these two on our mediterranean-habitat gradient, but to no more purpose. More to the point, we can refer doubting readers to yet another data set, the pigeons of mediterranean South Africa. Here, however, we census not just two species, but seven! They span together the same habitat range, but manifest an increased range of sizes (wings, 107 to 226 mm). This example demonstrates simultaneously that (1) convergent evolution does occur, for the earlier-cited similarities are put into perspective by the later-cited anomalies; (2) convergent evolution does not always occur (or is not always in evidence), since the South African habitats provide an exceptional mismatch in species numbers; but (3) convergent evolution may describe some aspect of the pigeon guild and not other aspects, for, although South African habitats have seven pigeon species in the mediterranean zone, five of them are quite rare. The two common species are *Streptopelia capicola* (wing, 157 mm) and *Columba arquatrix* (wing, 226 mm), a pair of species extremely similar to the three species pairs from the other three continents, the two common species from the South African seven providing, in fact, the closest size matches to mediterranean pigeons elsewhere.

In summary, our data must be interpreted in the light of the ranges of taxonomic, ecological, and morphological diversity that are known to exist in a particular taxon of consumers. Pigeons vary in size from as small as finches to as large as turkeys, and the relatively select group that comprises mediterranean species points up the relative similarity among the putative intercontinental counterparts.

What follows is first a discussion of the origins of the biota which confirms the low degree of genetic relationship between the plants and animals of Chile and California, a result that was an essential prerequisite of the experimental design. This is followed by an assessment of the climate and soils of the two regions, which shows their essential similarity, again fulfilling the research design requirements. The impact of man on these matched ecosystems is then assessed and found to be quite dissimilar in California and Chile, thus complexing our analysis to a certain degree. These introductory chapters are then followed by an analysis of the degree of convergence between the biotas, which was the principal objective of the study.

The Origin of the Biota

O. T. Solbrig
M. L. Cody
E. R. Fuentes
W. Glanz
J. H. Hunt
A. R. Moldenke

Ideally, our research design would have called for two regions with identical physical environments and with biotas that are completely unrelated phylogenetically. But no two physical environments, widely separated, can be identical, and no two areas of the world have biotas that are entirely unrelated. There are important similarities and contrasts, also, in the faunal and floral histories of mediterranean California and Chile, and in this chapter we shall look to the geological past to determine the character and the extent of phylogenetic relationship between the two modern biotas. Though we shall make reference to the faunas, the chapter will be concerned primarily with the two floras; the fossil record of the invertebrates is so poor that very little can be said regarding their history, and although the fossil record of the vertebrates is considerably better, they constitute only a small part of the biota as a whole.

PHYLOGENETIC RELATIONSHIPS
OF THE TWO FLORAS

In contrasting the two biotas, we must define first the levels at which the two biotas should be compared and the criteria upon which the comparisons should be made. For these purposes, the conventional hierarchical system of taxonomy is our most useful resource. Although taxonomic classifications are not phylogenetic in the strict sense, most taxonomists attempt to group together taxa that are similar and that appear to be related. Morphological similarity is the criterion most often employed, but where the available genetic data contradict the morphological evidence, most taxonomists tend to give priority to the genetic evidence. Where taxonomy is inadequate is in detecting convergences. In such cases, taxonomists are likely to group together phylogenetically unrelated but convergent taxa. Thus, if there is a bias in taxonomy, it is in the direction of exaggerating the actual phylogenetic relatedness of separate biotas, rather than the reverse. The extant taxonomic classifications are therefore well-suited for our purpose.

The exact degree of relationship of the mediterranean floras of California and Chile is difficult to assess, because of the lack of modern treatment of the flora of Chile. But by drawing upon data taken chiefly from Munz and Keck (1959), Reiche (1896-1911), Muñoz Pizarro (1966), Oberdorfer (1960), and Raven (1963), we can make some useful observations.

A visitor acquainted solely with the California flora will find, on first stepping on Chilean soil, a great many familiar plants, both cultivated and wild. The visitor—he might as easily be a Chilean in California—is likely to recognize among the wild plants most species of the ruderal vegetation and many of the herbaceous genera, but very few of the woody ones. Chile and California have exchanged many weeds and near weeds in the recent past: *Eschscholzia californica, Oxalis laxa, Nicotiana glauca, Madia sativa, Stipa brachychaeta,* among others. Both areas have been invaded by weeds introduced from the Mediterranean proper, such as *Bromus mollis, Erodium cicutarium,* and *Sonchus asper.* The woody shrubs and trees, however, are an entirely different story. Although some of the genera (*Haplopappus, Gutierrezia, Acacia, Larrea,* etc.) are represented in both North and South America, the species are not, and many of the genera and even some of the families are restricted entirely to either North or South America.

No species of shrubs or trees is shared between the mediterranean regions of Chile and California, and only very few genera, such as *Acacia, Alnus, Baccharis, Cassia, Condalia, Grindelia, Gutierrezia, Haplopappus, Krameria, Prosopis, Rhamnus, Schinus, Solidago* and *Verbesina.* Of these none is a dominant element of the vegetation. Some, such as *Schinus molle,* are introduced into both areas, while most of the other genera belong to phylogenetically advanced families (mostly Compositae) and are likely to be recent colonizers.

Table 5-2 lists the main species of shrubs and trees of the mediterranean regions of Chile and North America. All the species are indigenous. They have diverse affinities, however. Most of the genera of the mediterranean flora of California are derived from the Madro-Tertiary Geoflora which is of tropical derivation (Axelrod, 1958). However, some genera are derived from the Arcto-Tertiary Geoflora (Chaney, 1947; Axelrod, 1973). In Chile three main affinities can be recognized: taxa of tropical affinities, taxa of southern or Antarctic affinity, and taxa with affinities of the Andean flora. This last element evolved late in the Tertiary, and in the Pleistocene after the Andes reached their present height. Many of the Andean elements are North American migrants that invaded South America after the two continents were joined in late Pleistocene or shortly before.

Raven (1963) has discussed the relationship of the herbaceous flora of temperate South and North America. The number of species in common to mediterranean California and Chile is over 130. Most of these are autogamous plants from open habitats. However, the most abundant herbaceous species are European weeds, as a result of the high degree of man-made disturbance. A study of Gulmon (1977) illustrates this point.

Gulmon compared the herbaceous vegetation of central Chile with that of the San Diego area and that of northern California. A number of sites were chosen in each region, and at each site all species present at each of 50 disjunct 0.2 dm^2 sample quadrants were recorded. No species was recorded unless it was present in a sample, even if it was known to grow in the area. Eighty-seven species were observed at twenty sites sampled in the Santiago, Chile, area; eighty-five species were recorded at sixteen sites in northern California (eight sites in Palo Alto, eight in Hopland); and only fifty-three species were noted at thirteen sites in the San Diego area. The average number of species per 0.2 dm^2 quadrant was 4.2 in Chile, 4.6 in northern California, and only 2 in the San Diego area. The differences in species number are most likely due to differences in rainfall between the regions, but the Chilean region appears to be richer in species than comparable points in California.

Tables 2-1 and 2-2 list the species in all three regions that were present in at least 10 percent of the sample points. Only fourteen species were common to all three regions. Santiago and San Diego shared nineteen species; Santiago and northern California shared twenty-nine species. These figures include, however, all of the common species. Of the nineteen species common to Chile and southern California, seventeen are European weeds and the other two are native to California. Of the twenty-nine species shared by Chile and northern California, twenty-four are European in origin, one is a cosmopolitan weed, and four are native to the New World. Of these, *Madia sativa* is native to Chile and *Eschscholzia californica* is native to California; both are presumed to be recent invaders. At the generic level, Chile and California share approximately half of the herbaceous genera. Gulmon's study confirms that the similarity of the herbaceous vegetation derives chiefly from the large proportion of European weeds recently introduced by man.

EVOLUTION OF THE MEDITERRANEAN FLORAS
OF CALIFORNIA AND CHILE

As Raven (1973) has noted, the appearance of mediterranean floras must be viewed as part of the continuing evolution of subhumid to semiarid plant communities at the margins of the tropics. In western North America these communities have been termed the Madro-Tertiary Geoflora (Axelrod, 1958). They were formed primarily by elements derived from the warm, humid

TABLE 2-1   *Some Principal Herbaceous Species of the Mediterranean Formations of Central Chile, with Areas of Origin*

Species	Family	Native area
Aira caryophyllea	Gramineae	Europe
Bromus mollis	Gramineae	Europe
Bromus trinii	Gramineae	Chile
Erodium botrys	Geraniaceae	Europe
Erodium cicutarium	Geraniaceae	Europe
Erodium moschatum	Geraniaceae	Europe
Hypochoeris glabra	Compositae	Europe
Hypochoeris radicata	Compositae	Europe
Koelaria phleoides	Gramineae	Europe
Medicago falcata	Leguminosae	Europe
Trifolium glomeratum	Leguminosae	Europe
Vulpia bromoides	Gramineae	Europe

Source: From Gulmon (1977).

TABLE 2-2 *Some Principal Herbaceous Species of the Mediterranean Grassland Formations of California, with Areas of Origin*

Species	Family	Native area
Southern California		
Avena barbata	Gramineae	Europe
Avena fatua	Gramineae	Europe
Brachypodium distachyon	Gramineae	Europe
Bromus mollis	Gramineae	Europe
Bromus rigidus	Gramineae	Europe
Erodium botrys	Geraniaceae	Europe
Hordeum leporinum	Gramineae	Europe
Vulpia myuros	Gramineae	Europe
Northern California		
Aira caryophyllea	Gramineae	Europe
Avena fatua	Gramineae	Europe
Briza minor	Gramineae	Europe
Bromus mollis	Gramineae	Europe
Bromus rigidus	Gramineae	Europe
Festuca myuros	Gramineae	Europe
Linanthus bicolor	Polemoniaceae	California
Lolium multiflorum	Gramineae	Europe

Source: From Gulmon (1977).

Neotropical Tertiary Geoflora, but contained also elements derived from the cool, temperate Arcto-Tertiary Geoflora to the north (Chaney, 1936, 1947). Likewise in South America the mediterranean flora of Chile derives primarily from a cool-temperate flora formed by elements derived from the warm Neotropics, but containing also elements derived from the cool-temperate flora that once occupied the northern reaches of Antarctica, Australia, New Zealand, and the very southern tip of South America, during Cretaceous and perhaps early Tertiary time (Menendez, 1972). However, the characteristic narrowing of South America from the equator to the poles and the opposite situation in North America, as well as the timing of the mountain-forming events, had a decisive influence on the final distribution of tropical and temperate taxa in the mediterranean climatic areas of both continents. These events need to be considered in more detail separately for North and South America.

South America and Central Chile

This account will begin with the Cretaceous because it is the oldest period from which we have fossil records of angiosperms.

At the beginning of the Cretaceous, South America and Africa were probably still connected, and Antarctica with Australia (Dietz and Holden, 1970), since the rift that created the South Atlantic and separated South America and Africa apparently had its origin during the lower Cretaceous some 130 million years ago. The position of South America at this time was slightly south (approximately lat. 5° to 10° S) of its present position and with its southern extremity tilted eastward. There were no significant mountain chains at that time.

Although the first records of angiosperms date from the Cretaceous (Maestrichtian), the known fossil floras from the Cretaceous of South America are formed predominantly by pteridophytes, Bennettitales, and conifers (Menendez, 1969). Likewise, the fossil faunas are formed by dinosaurs and other reptilian groups. Toward the end of the Cretaceous (or beginning of Paleocene) appear the first mammals (Patterson and Pascual, 1972).

Climatologically, the record in the Cretaceous points to a much warmer and possibly wetter climate than today, although there is evidence of some aridity, particularly in the lower Cretaceous. In effect, the high rainfall in the present Amazonian region is the result of the condensation of moisture from rising tropical air that is cooling adiabatically. This air is brought in by the trade winds and acquires its moisture over the North and South Atlantic. Before the breakup of Gondwanaland, trade winds must have been considerably drier on the western edge of the continent after blowing over several thousand miles of hot land. It is interesting that some characteristic genera of semidesert regions, such as *Prosopis* and *Acacia*, are represented in both eastern Africa and South America. This disjunct distribution can be interpreted by assuming Cretaceous origin for these genera, with a more or less continuous Cretaceous distribution that was disrupted when the continents separated (Thorne, 1973; Raven and Axelrod, 1974). Finally, there is some geomorphological evidence for local aridity in the deposits of the Lower Cretaceous of Córdoba and San Luis in Argentina (Gordillo and Lencinas, 1972).

Paleobotanical evidence (Menendez, 1969, 1972) shows that the neotropical flora and the antarctic flora were distinct entities already in Cretaceous time, and that they have maintained that distinctiveness throughout the Tertiary and Quaternary in spite of changes in their ranges (mainly an expansion of the antarctic flora). The paleobotanical evidence futher suggests that at the beginning of the Tertiary the neotropical angiosperm flora covered all of South America, with the exception of the very southern tip.

During the Tertiary, South America and Africa drifted away from each other, while South America and North America, which were widely separated at the beginning of the period approached each other and eventually were joined when the Panama Isthmus was uplifted at the end of the Tertiary Period. Consequently, South America was an island continent during the Tertiary. Floristic and faunistic interchanges of a limited kind were possible with Africa at the beginning of the period and with North America toward the end.

As far as can be ascertained, at the beginning of the Tertiary there were no large mountain chains in South America. The period was characterized by the gradual uplifting of the Andean chain, which became accelerated after the Eocene and culminated in the Pliocene and Pleistocene. Climatically, the Tertiary in South America was characterized by the gradual cooling and drying of the climate after the Eocene, apparently a worldwide event (Wolfe and Barghoorn, 1960; Axelrod and Bailey, 1969; Wolfe, 1971), with a temperature decrease of about 7° to 10° C from Paleocene to Pliocene.

At the beginning of the Tertiary the flora of South America was predominantly tropical and subtropical. However, it had acquired a character of its own, very distinct from contemporaneous European floras, although there were resemblances to the African flora (Van der Hammen, 1973). This flora reached its greatest extent during the Eocene, as shown by the fossil flora of Rio Turbio in Argentine Patagonia. Here the lowermost beds containing *Nothofagus* fossils are replaced by a rich flora of tropical elements with species of *Myrica, Persea, Psidium*, and others, which is then again replaced in still higher beds by a *Nothofagus* flora of more mesic character (Hünicken, 1966; Menendez, 1972; Romero, 1973). Paleobotanical evidence (Menendez, 1969, 1972) shows this tropical flora to have been mostly evergreen, but there is reason to believe (Solbrig, 1976) that some more dry adapted elements prevailed at middle latitudes.

The early Tertiary mammalian fossil faunas consist of marsupials, edentates of the suborder Xenarthra, and a variety of ungulates (Patterson and Pascual, 1972). These forms appear to have lived in a forested environment, confirming the paleobotanical evidence.

In the Eocene, evidence of a more open, drier vegetation, particularly grasses can be found (Menendez, 1972). This was also a time of radiation of several mammalian phyletic lines, particularly marsupials, xenarthrans, ungulates, and notoungulates (Patterson and Pascual, 1972). More interesting is the acquisition of high-crowned, or hyposodont, and of rootless, or hypeseledont teeth by certain ungulate lines (Patterson and Pascual, 1972). Such teeth are usually associated with a herbaceous diet, particularly one of grasses. By the Lower Oligocene such teeth had been acquired by no fewer than six groups of ungulates. Such animals must have thrived in the evolving grassland areas. True grasslands such as the Argentine pampas are probably younger, but by the Eocene it seems reasonable to propose the existence of open savanna woodlands, somewhat like the llanos of Venezuela, or the Brazilian "cerrados" today. During the Oligocene the caviomorph rodents and platyrrhine primates appear, which probably arrived from North America via a sweepstakes route (Simpson, 1950; Patterson and Pascual, 1972), although an African origin has also been proposed (Hoffstetter, 1972).

The gradual deterioration of the climate and the uplifting of the Andean chain and adjacent mountains, especially after the Eocene, had three principal effects on the biota: (1) the gradual retreat of the tropical woody flora towards northern latitudes and the advance of the south temperate flora which reached its present extent by Pliocene time; (2) the evolution of dry-adapted communities in midlatitudes, such as grasslands (from Eocene or Oligocene time on) and semideserts (in the Oligocene or the Pliocene); and (3) the evolution of a mountain flora in the gradually uplifting Andes. The taxa of the dry-adapted and mountain floras are derived from tropical ancestors, with some admixture of elements derived from the south-temperature flora and from North America, such as *Alnus, Sambucus,* and members of the Cruciferae, Compositae, Gramineae, etc. In the Pliocene, after the two continents joined at the isthmus, the North American elements became more numerous (Graham, 1973).

As far as can be ascertained, Chile at the end of the Pliocene possessed a flora formed from the following vegetation types: (1) dry-adapted, tropical formations in the north and center; (2) cold-humid forest of a south-temperate origin in the south; and (3) typical Andean elements in the mountains, these deriving chiefly from tropical ancestors which were highly modified by the Pliocene.

The Pleistocene in South America had drastic effects on the vegetation, especially in Chile. In South America, glaciation took the form chiefly of mountain glaciers. There is evidence (Vuilleumier, 1971) that they reached the Chilean coast along some of the major river valleys, such as the Maipo. At least three glacial events have been recorded for Chile, of which the Würm was the most drastic.

The landscape in southern South America differs markedly from what it was at the time of the glacial maximum, about 20,000 years ago. A greatly expanded Patagonian ice field covered virtually all of the region south of Valdivia, and valley glaciers extended eastward from the Andean crest into Patagonia (Flint, 1971). Nearly the whole area now covered by broad-leaved forests in southern South America, including much of the southern central valley of Chile, was covered by ice, treeless steppe, or tundra vegetation (Heusser, 1974). From pollen data, Heusser estimates that mean summer (January) temperatures were 8° C lower than recent normal values in south-central Chile. A temperature depression of 6 to 8° C is also postulated for equatorial localities in Colombia (van der Hammen and Gonzales, 1960).

Assuming an equatorward shift in rainfall paralleling the shift in isotherms, the areas of central Chile now supporting chaparral, savanna, and desert scrub vegetation were probably covered by forests like those found in southern Chile today. The southern elements in the vegetation of the "fog forests" at Fray Jorge and other localities along the central coast (Muñoz Pizarro and Pisano Valdés, 1947) could be relicts of the last glacial period. The last major glacial readvance in southern Chile ended about 14,000 years ago (Mercer, 1972). Essentially modern vegetation patterns were then reestablished. The warming trend continued, culminating about 6,500 years ago when temperatures may have been 1 to 2° C above recent norms; this warm period may have been associated with increased seasonal dryness. Cooler and possibly wetter conditions have prevailed during the past few thousand years (Heusser, 1974). In general terms, during the cold-humid periods the southern *Nothofagus* communities moved north, whereas during the dry, warm interglacials the dry, tropical, and semitropical communities expanded. At the same time, the high-mountain vegetation expanded and withdrew with the ice. The pluvial regime also changed (if it had not already in the Pliocene or earlier), especially in central Chile: summers turned dry, and winters wet, bringing on the present mediterranean climate. The advances and retreats of the various vegetational elements, the isolation of the region as a result of glaciation and glacial lakes, and the marked climatic changes led to the rapid evolution of what we call today the mediterranean flora of Chile, mostly by elimination of elements that were not able to withstand summer drought. It is not surprising that the flora is formed by elements derived from the three types of vegetation discussed, nor is it surprising that these elements tend to predominate today in communities having a resemblance to their ancestral habitat: the dry, subtropical elements (*Hoffmannseggia, Cassia, Acacia, Baccharis,* Cactaceae, Zygophyllaceae, etc.) in communities that are dry, particularly in the Norte Chico; the Andean element (Nolanalceae, Malesherbiaceae, *Astragalus, Ribes, Sanicula, Saxifraga*) primarily in areas with cold winters; and the southern element (Lauraceae, *Drimys*) in the wetter communities).

North America and Central California

At the beginning of the Tertiary the vegetation in North America was much like that in South America (see the discussion above): to the north was the Arcto-Tertiary Paleoflora (Chaney, 1947); whereas a vegetation of tropically derived plants known as the Madro-Tertiary Geoflora (Axelrod, 1958) occupied the rest of western North America. Both of these floras furnished elements of what we now call the "mediterranean vegetation." Because of the superior North American fossil record, the evolution of plant communities can be described in more detail and more accurately for California than for Chile. But the broad outlines of the history of the vegetation appear to be similar in the two hemispheres: a gradual deterioration of the climate, with a concomitant advance of polar elements; a gradual increase of dry-adapted elements, with an indication of semidesert vegetation from the Oligocene to the Pliocene on; and an increase in orogenic activity, which had its culmination in Pliocene and Quaternary times. Pleistocene events in turn affected the flora, and were directly responsible for the emergence of mediterranean climates as we know them today.

Axelrod (1973, 1975), in reviewing the history of the mediterranean ecosystems of California, divides the vegetation into four major types: (1) mixed evergreen forest (tan oak-madrone), (2) oak-laurel forest, (3) oak woodland-savanna, and (4) chaparral. He shows that all four formations evolved before the appearance of a typical mediterranean climate (summer drought, winter rain), and that they extend today into areas with summer rain. By Miocene times all four vegetational types had emerged. Furthermore, most of the woody species of these formations were already present in the Miocene or the Pliocene. Not so the herbaceous flora, which tends to be restricted to the mediterranean climatic area today, and presumably evolved recently as the direct result of the area's having acquired a regime of winter rains and summer drought. This shift was brought on by orogenic activity, and by the appearance of a cold ocean current off the coast of California. However, the woody assemblage belonging to the four major communities was much richer in taxa prior to the Pleistocene, and the main effect of the appearance of a mediterranean climate was the elimination of a number of these taxa. Nonetheless, a few woody taxa, especially the genera *Ceanothus* and *Arctostaphylos*, have undergone rapid speciation and the formation of a number of endemic populations (Nobs, 1963; Wells, 1969; Stebbins and Major, 1965; Axelrod 1973).

The late-glacial and postglacial climatic history of California is generally similar to that of Chile. Comparable decreases in temperature during full-glacial times are indicated by the depression of vegetation zones in the Mohave Desert just inland from the modern coastal belt of Mediterranean scrub vegetation (Wells and Jorgensen, 1964; Mehringer, 1965; Wells and Berger, 1967; Mehringer and Ferguson, 1969). Increased rainfall and reduced evaporation are shown by the filling of closed desert basins with deep lakes (Smith et al. 1967), and by the thick valley glaciers that flowed down from the ice-capped Sierra Nevada. Although deglaciation may have lagged behind that in the Southern Hemisphere, some high-altitude basins were ice free shortly after 10,000 years ago (Adam, 1967). A postglacial thermal maximum perhaps 6000 to 7000 years ago, followed by cooling, is shown by pollen evidence (Adam, 1967); by treeline history (LaMarche, 1973), and by long tree-ring records (LaMarche, 1974).

Common Factors in the Evolution of the Two Floras

The evolutionary sequence in western North and South America is thus fairly similar. In Cretaceous times each area was settled by a tropical-to-subtropical flora. As the climate slowly deteriorated during the Cenozoic, elements of the floras adaptable to increasing drought were selected. Furthermore, the emergence of sizable mountain chains in what had been, in each case, rather even terrain created a new environment, and taxa evolved to occupy the new habitats. The decline in temperature determined an equatorward migration of elements of a distinctly cold-adapted vegetation that had evolved in both hemispheres at high latitudes. These elements tended to occupy the colder and wetter habitats. In both areas, then, the Pleistocene had drastic and lasting effects, producing the present "mediterranean" vegetation. Its principal effect was the alteration of the climate: the formation of the polar ice caps and the resulting cold currents created the cool, dry summers and mild, somewhat wet winters typical of the regions. The evolutionary effect of this climatic regime was to impoverish the woody flora by eliminating a number of its elements (Axelrod, 1973). A second effect of the Pleistocene was the creation, in both regions, of isolated pockets where rapid speciation could take place. The increased orogenic activity in both the Sierra Nevada-Cascade chain of western North America and the Andes of South America also contributed to rapid speciation. Finally, in recent time, the two areas have exchanged herbaceous taxa.

PHYLOGENETIC RELATIONSHIPS OF THE TWO FAUNAS

The phylogenetic relationship between the faunas of mediterranean Chile and California is even more tenuous than that between the floras. With the exception of certain animals introduced by man, such as the honeybee, only large, far-ranging carnivores like the puma are common to both areas.

There is essentially no fossil record of invertebrates in South America. And because the taxonomic knowledge of the present-day invertebrate fauna is very inadequate—still at the alpha stage in most phyla—it is not possible to reconstruct the history of these groups.

The fossil record of South American vertebrates, not including fishes, is better, but only the mammals record is perhaps adequate. The evolutionary history of South America terrestrial vertebrates has proceeded in isolation from that of North American vertebrates, at least throughout most of the Cretaceous and Tertiary. Since this coincided with the time of radiation of the major groups of mammals, it had a pronounced effect on the two modern mammal faunas. In general, South America is characterized during the Tertiary by the radiation of marsupial groups that were absent in North America, as well as certain groups of placentals, such as notoungulates, whereas in North America the placental mammals were uniformly dominant, including groups such as the carnivores that did not exist in South America during the Tertiary. Following the connection of the two continents at the end of the Pliocene, extensive faunistic interchanges took place, as well as the extinction of some South American groups, notably the notungulates and the marsupial carnivores (Patterson and Pascual, 1969).

Unfortunately, none of the major fossil beds are situated in the mediterranean region of Chile, and the evolution of the fauna, therefore, cannot be documented. However, the floristic changes discussed above probably determined the character of the fauna of these regions.

We now proceed to enumerate the phylogenetic similarities and differences in the five animal taxa studied in detail: ants, bees, lizards, birds, and mammals.

Ants. The ant communities of Chile and California were studied at four pairs of matched sites; in California the sites yielded a total of fifty species, only twenty-two species were found at the Chilean sites. The Californian species represent twenty-one genera, of which five are shared with Chile and were found to be common; another (*Conomyrma*) has a close Chilean counterpart (*Araucomyrmex*); twelve of the remaining fifteen had no representatives in Chile, and ten of the remaining fifteen were encountered only rarely in California. While five of the Chilean genera were shared with California and another was closely matched, the remaining four are South American endemics.

Bees. Bee speciation has been extensive in both the southwestern United States and Chile (as well as in other semiarid areas of the world); on a bee species/area basis these two areas may be, respectively, the richest and second richest regions in the world.

The considerable phylogenetic similarities between the ant faunas of California and Chile do not extend to the bee faunas. At the level of the species, the faunas of the two continents are extremely distinct; they share only the cosmopolitan honey bee *Apis mellifera*. Table 2-3 summarizes the species distribution in California and Chile among higher taxa. At the family level, both regions have representatives of six families that have nearly worldwide distributions; in addition, both California (with Mellitidae) and Chile (with Fideliidae) support one additional but numerically unimportant family. Furthermore, most of the bee subfamilies and tribes occur in both California and Chile, and even at the genus level there are many cosmopolitan taxa (e.g., *Anthidium, Megachile, Xylocopa, Bombus, Centris, Colletes*) that are common to both regions. Where common North American groups such as Andreninae, Hylaeini, Dufoureinae are virtually absent in Chile, their places are filled to some extent by the very diverse Chilean Colletidae fauna.

Lizards. In contrast to the ants and to the insects in general, the lizards of mediterranean Chile and California are taxonomically quite distinct (Doñoso-Barros, 1966; Stebbins, 1966). Chile is celebrated for the tremendous radiation of species within one genus, *Liolaemus*, and seven of the nine Chilean lizard species encountered on three sites were this genus (family Iguanidae). Californian species are far more diverse taxonomically, and the eight species found on three matching sites comprise six genera and four different families (Iguanidae, Teidae, Anguidae, and Scincidae).

Birds. The taxonomy of birds of mediterranean California and Chile has been worked out with considerably more detail than has lizard or bee taxonomy, but the picture is the same. The two faunas are quite distinct, and, although there are well represented families in common between the two regions, such as the flycatchers Tyrannidae and the finches and buntings Fringillidae and Emberizidae, a good proportion of the common birds on one

TABLE 2-3 *Californian and Chilean Bee Taxa Compared, by Species Numbers and Flower-Feeding Habits (s = specialist; g = generalist; fac - facultative)*

Taxon	Region	
	California	Chile
COLLETIDAE		
Paracolletini	0	28 nearly all s
Colletini	41 most s	47 most s
Xeromelissinae	0	39 probably most s
Hylaeinae	28 all g	0
Diphaglossinae	0	14 most g
ANDRENIDAE		
Andreninae	294 nearly all s	3* probably all s
Panurginae	325 nearly all s	64 nearly all s
HALICTIDAE		
Halictinae	301 all polylectic	77 all polylectic
Nomiinae	5 g/fac. s	0
Dufoureinae	79 nearly all s	2 s
MELITTIDAE	23	0
FIDELIIDAE	0	2 unknown
MEGACHILIDAE		
Lithurginae	1 s on cactus	6 s on cactus
Anthidiini	77 mostly s on composites, legumes	36 s on legumes
Megachilini	326 mostly s on composites, legumes, borages	70 mostly s on composites, legumes
APIDAE		
Apini	1 social g	1 social g
Bombini	37 social g	3 social g
ANTHOPHORIDAE		
Xylocopinae	18 g	6 g
Anthophorinae		
solitary	205 g or s	62 g or s
parasitic	212 g	29 g
Total Species	1,980	485
Total Genera	84	66
Genera Shared	20 (24%)	20 (30%)
Genera Not Shared	64 (76%)	46 (70%)
Species in Shared Genera	1,253 (63%)	152 (32%)
Species Not in Shared Genera	771 (37%)**	266 (68%)

Source: Moldenke (1976).

*Tenuous subfamilial placement.

**Approximation because the very large holarctic genera *Andrena, Nomada,* and *Osmia* are in need of considerable revision.

continent belong to families not present at all on the other. The species censused in California belong to twenty-one families, those in Chile to sixteen, and ten of these families are shared between continents. In particular, many common Chilean birds belong to the ovenbird and tapaculo families Furnariidae and Rhinocryptidae, and in California many wood warblers (Parulidae), vireos (Vireonidae), and titmice (Paridae) were found, all families unrepresented in censuses from the other continent.

Mammals. Our studies of mammal species were confined to species smaller than a cottontail rabbit, and almost all of the species found were rodents (Rodentia): nine of the ten species totaled in Chile, sixteen of the eighteen tallied in California. In addition, a marsupial, *Marmosa elegans*, occurred in Chile, and two insectivora, both shrews, were found in California. But, although most of these species fall into the same order, Rodentia, just over half of the species from each continent belong to the only two families they share, Cricetidae (four Chilean and eight Californian species and Muridae (one species in each region, both introduced), and five of the nine families present occur in just one or the other country. In fact, no genera are held in common between the two species lists (Glanz, 1975), and the mammal faunas are judged to be phylogenetically distinct.

SUMMARY AND CONCLUSIONS

The phylogenetic relationship of the biotas of the two areas is low. At the species level endemism in both regions is relatively high. At the generic and family levels there are few common taxa among the woody elements of the flora, but a large number among the herbaceous ones. Most of the common species have migrated into one or the other region very recently because the areas share a similar environment. As to the origin of the biota, among the plants they are mostly of tropical origins in both areas with some cool-temperate elements. Little is known regarding the invertebrates. As to the vertebrates, extensive exchanges have taken place. The amphibians and reptiles appear to be mostly of South American origin, the mammals of North American origin and the birds of tropical ancestry, both North and South American.

Past and Present Environment

P. C. Miller
D. E. Bradbury
E. Hajek
V. LaMarche
N. J. W. Thrower

27

The environmental bases of biotic similarities between widely separated ecosystems are their similarities in climate, geology, and soils (Figure 3-1), which are included here in the term macroenvironment. Given similar macroenvironments and sufficient time for evolution to occur, the result could then be a convergence in the structure of ecosystems and of organisms of different phylogenetic stock. This chapter documents the similarity in macroenvironments of the mediterranean scrub regions of California and Chile, a similarity evidently resulting in one case of convergent evolution.

THE REGIONS AND THE RESEARCH SITES

Owing to the global atmospheric circulation patterns and ocean currents, mediterranean scrub regions of the world center at about 33° latitude north and south of the equator, chiefly on the west sides of continents. Five such regions can be identified: around the Mediterranean Sea, on the southwest coast of South Africa, in southwestern and southern Australia, in southern California, and in central Chile (Figure 3-2). These regions exhibit similar generalized climatic characteristics that include warm or hot, dry summers with high solar irradiance and high rates of evaporation and mild, yet winters with low solar irradiance and low rates of evaporation (Figure 3-3; Table 3-1). Most of the precipitation occurs in a few storms of high intensity during the cooler period of the year. The storm and runoff patterns of this climatic regime lead to similar landscapes wherever these conditions occur.

Of the five mediterranean scrub regions of the world, the regions of California and Chile are particularly well suited for comparison, because their climates and geography are remarkably similar (Aschmann, 1973; di Castri, 1973; Thrower and Bradbury, 1973). Each country is marked by a coastal plain on the west and a mountain range not far inland (Figure 3-4). In southern California the coastal plain extends inland on the average for about 20 km. There the foothills begin rising gradually from about 100 m elevation, and at about 70 km inland the mountain summits reach 1,500-2,000 m

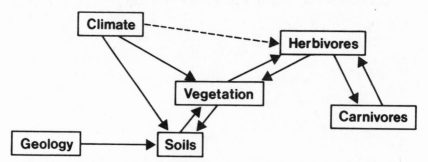

FIGURE 3-1. Interrelationships between climate, geology, soil, vegetation, herbivores, and carnivores. *(A dashed line implies a lesser influence.)*

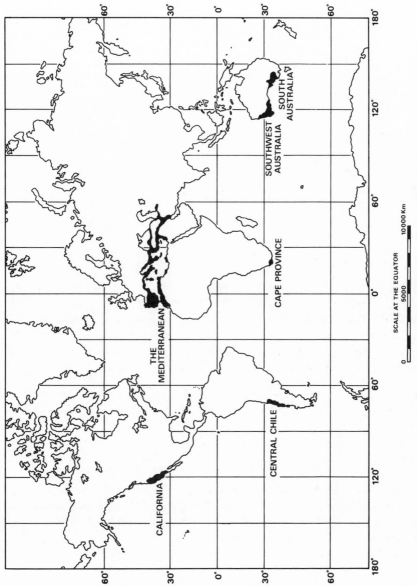

FIGURE 3-2. *World map of areas with mediterranean climates (from Thrower and Bradbury, 1977).*

29

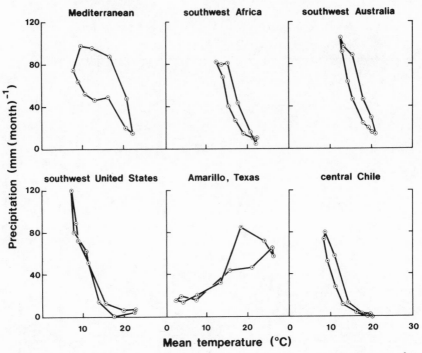

FIGURE 3-3. *Seasonal progression of monthly mean air temperature (°C) and monthly precipitation (mm) in the five regions of the world with mediterranean-type climates, and at Amarillo, Texas, an area with similar annual precipitation but summer rainfall and grassland vegetation.*

elevation. River valleys from coast to mountain tend to be tortuous, but neither deeply nor widely dissected. East of the mountains the land descends abruptly to about 300 m elevation and continues at this elevation, with isolated mountain ranges, for 600 km. In central Chile the coastal plain extends about 40 km inland, where the coastal range begins to rise abruptly, reaching 2,000 m elevation about 65 km inland. The land then descends eastward into a central valley about 20 km wide, which is flanked on the east by the Andes, rising to 6,000-7,000 m elevation. River valleys descending from the Andes, notably those of the Rio Aconcagua and Rio Maipo, tend to be straight and broad.

To document the degree of similarity in macroenvironments, the climate and soils were measured in detail at coastal, inland, and mountain research stations in the two countries (see Figures 1-1 and 1-2). In Calfornia climates and soils were measured at Torrey Pines State Park and at Camp Pendleton Marine Corps Base, which are located along the coast about 20 km and 80 km north of San Diego, respectively. Climate, soils, and microclimates were measured at Echo Valley, about 6 km north of Descanso on an east-facing slope in the higher foothills of the coastal mountains. A mountain station was

TABLE 3-1 *Summary of Climatic Parameters of Mediterranean-type Climates in Different Regions of the World**

Regions	Elevation (m)	Latitude	Global radiation (cal/cm² day)		Mean temperatures (°C)					Annual precipitation (mm)	Evaporation (mm/mo)		References
			Summer	Winter	Summer Max	Summer Min	Winter Max	Winter Min	Annual		Summer	Winter	
Mediterranean region													
Seville (southern Spain)	30	37°24'	558	174	36	20	15	6	19	535	285	68	Artéry (1970)
Almeria (southern Spain)	7	36°50'	596	215	29	21	16	8	18	231	93	56	"
Nice (southern France)	5	43°40'	612	154	27	19	12	3	15	862	134	–	Linés Escardo (1970)
Ajaccio (Corsica)	4	41°55'	638	142	28	16	13	3	15	672	132	–	"
Fez (Morocco)	415	34°02'	630	–	37	17	16	5	18	554	221	78	Wallén & Brichambaut (1962)
Tunis (Tunisia)	3	36°50'	–	194	32	20	15	7	19	461	–	–	Griffiths (1972)
Casablanca (Morocco)	49	33°34'	–	–	26	19	17	7	18	511	–	–	"
Jenin (Jordan)	138	32°28'	–	–	33	21	17	9	20	491	–	–	Wallén & Brichambaut (1962)
Southwestern Africa													
Capetown	17	33°54'S	732	241	26	16	17	7	17	506	322	71	Schulze (1972)
Langgewens	91	33°17'S	–	–	21	17	18	8	17	368	–	–	Wallén & Brichambaut (1962)
Southwestern Australia													
Esperance	4	33°50'	–	–	25	16	17	8	16	679	234	46	Gentilli (1971)
Perth	60	31°57'	650	235	29	16	17	9	16	889	263	45	"
Eyre	5	32°14'	–	–	26	15	17	6	18	289	–	–	"
Central Chile													
Banos de Jahuel	1,130	32°51'	678	161	31	15	16	6	16	345	–	–	Oficinia Meteorologia de Chile (1930–70)
Fundo Santa Laura	1,000	33°04'			23	11	12	4	16	1019*	–	–	"
Los Andes	816	32°50'			32	12	17	3	14	319	–	–	"
Santiago	520	33°27'			30	12	15	4	14	344	–	–	"
Zapallar	30	32°33'			23	14	14	8	14	356	–	–	"
Valparaiso	41	33°01'			23	13	16	8	10	512	–	–	"
Southwestern United States													
Cuyamaca	1,425	32°59'	649	270	30	13	10	2	12	–	–	–	NOAA (1970–74)
Julian	1,114	33°50'			32	11	13	1	13	604	–	–	"
Echo Valley	1,070	32°54'			32	11	15	1	17	438	–	–	"
Descanso	1,066	32°51'			–	–	–	–	–	–	–	–	"
Henshaw	822	33°14'			33	12	15	–2	14	536	–	–	"
Alpine	528	32°44'			33	15	18	5	17	362	–	–	"
Oceanside	25	33°12'			23	17	17	6	16	226	–	–	"
San Diego	5	32°44'			24	18	18	9	17	250	–	–	"

*This average covers a three-year period and includes one unusually wet year; the average is about 550–600 mm per year.

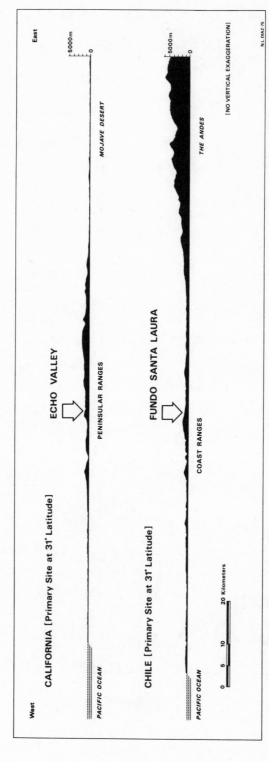

FIGURE 3-4. *Diagrammatic transect from the coast to the interior at latitude 32.5° in California and Chile (from Thrower and Brad-bury, 1977).*

32

maintained at the San Diego State University Observatory on Mt. Laguna. In Chile, climates and soils were measured near Cachagua on the coast and climate, soils, and microclimate were measured at a primary site at Fundo Santa Laura, inland on the east flank of the coastal range. No climatic station was established in the mountains of Chile because of security problems when the project began. In both countries the primary sites are 60–65 km from the coast.

The two primary sites are located in areas of similar tectonic history, and they share certain other common characteristics; for example, both are on generally east-facing slopes dissected by drainage to the east, both are of similar distance from the coast, and both are at a similar elevation. However, there are also some differences in their regional topographic position. Echo Valley is situated in a relatively shallow basin within the massive gradually ascending, tabular upland of the west slope of the Peninsula Ranges of southern California. Fundo Santa Laura is on a regionally east-facing slope within a series of ridges that form the broken crest of the Coast Range system of Chile. The presence of a massive cordillera (the Andes) in Chile and the lack of such a feature in the California area represent a substantial difference in these two sites—a difference that influences continental air masses affecting the two regions.

The studies summarized here involved syntheses of existing weather records from established stations, collection of climatic and geological data from the coastal and primary (inland) research sites in the two countries, and detailed analyses of the microclimate and soils at the primary sites. In each country the climate and vegetation changes from the coast inland. The hypothesis, for which the measurements were made, was that the climates of the inland areas in the two countries would be more similar to each other than the climates of the inland and coastal areas within each country. Similarily, the climates of the coastal areas of the two countries should be more similar than the climates of the coastal and inland areas within each country. Microclimatic information was collected at the primary sites in order to relate the macroclimatic data to the structure and function of the species, via the physical processes of energy exchange. Geological and soils data were collected to document the degree of similarity of the coastal and primary research sites.

PAST CLIMATIC REGIMES

Climate varies on time scales ranging from millions of years to a few decades or less. Broad global climatic trends, such as the general cooling that took place during the Tertiary and the alternating glacial and interglacial climates of the later Quaternary, have been important influences on the development of modern ecosystems. The distribution of plants and animals in the midlatitudes of both hemispheres changed greatly between glacial and interglacial times during the Pleistocene. Under full-glacial conditions, characterized by global cooling, lowered sea levels, and extensive continental glaciation, life zones were displaced far downward or equatorward of their interglacial positions. As Axelrod (1973) has emphasized, the interaction of such climatic fluctuations with topographic changes during the Pleistocene has contributed greatly to species diversity in regions of mediterranean climate (see Chapter 2).

Although many paleoclimatic indicators, such as fossil pollen, can indicate the direction and magnitude of past climatic fluctuations, their records usually can permit only general conclusions about past climatic variability. More detailed records for shorter and more recent time periods can often be obtained from the annual rings of trees. Trees growing in habitats where climate is frequently limiting to physiological processes can show large year-to-year differences in ring width that are closely linked to departures from climatic norms. Such trees are typically found in marginal locations, close to climatically determined limits of distribution. At 32° latitude, trees located toward the dry equatorward limits of the mediterranean climatic zone, or near the lower limits of forests in the mountains of this zone, may contain ring width records sensitive to moisture fluctuations. If the tree-ring variations are in fact highly correlated with climate, then the tree-ring record can be used to extend the meteorological record several hundred years into the past.

In southern California trees of several coniferous species attain great age and are known to yield moisture-sensitive records. Because the meteorological record for San Diego is among the longest available from the mediterranean zone in California, it can be used to characterize climatic variations that have occurred since the beginning of instrumental observations, about 125 years ago. These climatic records can be extended by correlating the records with tree-ring data and using the tree rings to estimate previous climatic history.

The longest tree-ring chronology that is reasonably close to San Diego comes from bigcone spruce (*Pseudotsuga macrocapa*) in the southern San Jacinto Mountains, about 100 km north of San Diego. Ring-width indices in a number of trees (Drew, 1972), representing the mean annual growth occurring in late spring and early summer, are significantly correlated with the total precipitation from the preceding July through June. Using weights calculated by regression analysis, annual precipitation amounts were estimated from ring-width indices from A.D. 1458, the first year of the tree-ring record, to 1850, when precipitation measurements began (Figure 3-5).

A similar analysis was carried out for central Chile using the precipitation from January through December. The tree-ring data were obtained from an isolated stand of ciprés (*Austrocedrus chilensis*) representing the northernmost known occurrence of this species (Schlegel, 1962). This stand, in the coastal range near San Felipe, is about 100 km north of Santiago. A tree-ring chronology beginning in A.D. 1010 was developed using samples from a large number of trees. The ring-width indices are highly correlated with annual precipitation in Santiago. Using the tree-ring data, estimates of annual precipi-

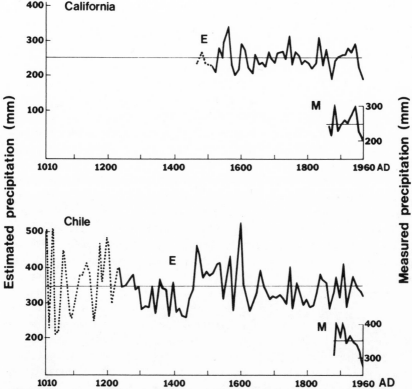

FIGURE 3-5. *Decadal means of annual precipitation (mm) measured (M) at San Diego, California, and Santiago, Chile, plotted at the first year of the decade. Estimated (E) values are based on tree-ring data; lines are broken where small sample size has increased uncertainty of estimates. Straight lines are means for entire time span. The year is July-June for San Diego and January-December for Santiago.*

tation were made from A.D. 1010 to 1866, when the continuous observational precipitation record begins.

Spectral analysis (Jenkins and Watts, 1968) shows how the total variation in a climatic record is distributed into climatic fluctuations of different lengths or frequencies. For both San Diego and Santiago, spectral analysis was used to compare the correlation between the precipitation estimated from tree rings and the precipitation recorded in the climatic record. In both regions the precipitation histories estimated from tree-ring data have less variation at high frequencies (short-duration climatic fluctuations) than do the precipitation histories recorded in the climatic data. However, for climatic fluctuations with periods of more than ten years (i.e., with frequencies less than 0.1 cycles per year) the variations in climatic data and tree-ring estimates are comparable. The coherency between the climatic data record and the tree-ring estimate from each city is also highest with low frequency climatic fluctuation (long-duration climatic changes). Thus, precipitation amounts estimated from tree-ring data appear to faithfully record the lower frequency or long-term fluctuations in climate.

Climatic variability in the mediterranean zones of California and Chile can now be compared not only for the period of the meteorological record, but also for the past several hundred years. A comparison of the relative frequency of variation with different lengths or periods (power spectra) in the recorded annual precipitation for the two regions shows that the variation in annual precipitation is less in California than in Chile at all but the shortest periods. Variation in rainfall from one year to the next (one-half cycle per year) is greater in California. The spectra of precipitation for the past several hundred years estimated from tree rings also indicates the difference in low frequency variations. California shows less low frequency variation (climatic fluctuations of long duration) than Chile. Moreover, the variations recorded during the past 125 years at San Diego seem representative of the kind that have occurred during at least the past 500 years. In Chile, on the other hand, the meteorological record seems a poor guide to kinds of variations that have taken place in the geologically recent past. Rainfall in central Chile has tended to remain much above or much below the long-term average for periods of a hundred years or more. In southern California the overall range is smaller and periods of drought or deluge have rarely persisted for more than thirty years.

The persistence of climatic anomalies from one year to the next was measured by conditional probability analysis, using the annual precipitation records for San Diego and Santiago (Figure 3-6). The analysis indicated that an unusually wet year in either city is likely to be followed by a year with near average rainfall, but an unusually dry year will probably be followed by another dry year. This tendency is more pronounced in Santiago, where the rainfall in years following dry years was below average or much below average in two-thirds of the cases. The year-to-year correlations indicated in the probability analyses are consistent with the results of the spectral analysis in showing that rainfall in San Diego has greater year-to-year variability and less tendency for persistence or year-to-year climatic similarity.

Some general conclusions of potential biological significance can be drawn from the evidence for comparative climatic variability in the mediterranean climatic zones of California and Chile. On a time scale of 10^3 to 10^6 years, corresponding to the length of time required for large-scale movement of life zones and for the operation of evolutionary processes, climatic changes in California and Chile have probably been similar in direction, magnitude, and timing. For time periods of 10 to 10^3 years (which encompass the life spans of most shrubs and trees) climate, as measured by precipitation, has been less variable in California than in Chile. On a time scale of one to a few years (which encompass the lifetimes of annuals and biennials and the fruiting cycles of most other plants) precipitation shows greater year-to-year variabil-

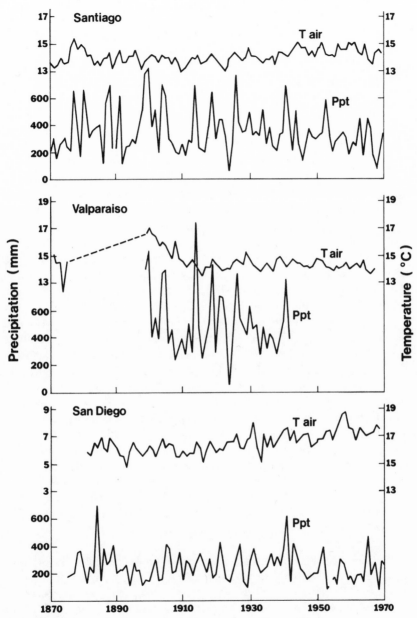

FIGURE 3-6. *Annual mean air temperatures (°C) and precipitation (mm) for San Diego, California, and Santiago and Valparaiso, Chile.*

ity in California than in Chile, and there is less tendency for successive drought years to occur in California than in Chile.

PRESENT CLIMATIC REGIMES

Synoptic Climatology

The climates in the mediterranean regions of southern California and central Chile are dominated by subtropical high pressure cells offshore and cold ocean currents. These areas of high atmospheric pressure inhibit the passage of frontal storms across the mediterranean scrub regions and also increase the vertical stability of the air. In the summers in both regions, the areas of high pressure are located away from the equator and frontal activity is displaced towards the poles, whereas in the winters the high-pressure cells move closer to the equator, and frontal storms pass closer to the equator, crossing the mediterranean scrub regions. These winter frontal storms are the main source of precipitation in both regions. Since the latitude of the high-pressure cell is about $5°$ closer to the equator in the Southern Hemisphere (varying between about $2°$ and $7°$), polar fronts in the Southern Hemisphere penetrate closer to the equator (van Loon, 1972). Thus, at the same latitude precipitation is lower in California than in Chile (di Castri, 1973). In both countries precipitation increases inland because of orographic lifting of the air as it moves across the foothills and mountains.

Atmospheric stability reduces the frequency of convectional storms and contributes to an inversion layer caused by the penetration of cooler marine air under the warmer upper air. In southern California the inversion layer averages about 400 m deep; in Chile it appears to be 400–700 m deep and more stable (Figure 3-7) (Fuenzalida Ponce, 1967; Schneider, 1969). Unfortunately, most of the established climatic stations in both countries are located within the inversion layer and it becomes difficult to generalize about the climate of the mountainous regions that occur above the inversion, especially in Chile.

Regional patterns of air movement create similar precipitation and temperature regimes in the mediterranean regions of the two countries (Figures 3-8, 3-9). Precipitation is between 200 and 800 mm annually, occurring predominantly in the winter; precipitation increases from the coast to the interior with increasing elevation. From the centers of the two regions it increases towards the respective poles and decreases toward the equator. In California precipitation increases from about 250 mm at San Diego ($32°30'N$) to about 995 mm at Fort Bragg ($39°24'N$), a rate of change of 108 mm per degree of latitude, and decreases to about 100 mm at San Quintin ($30°30'N$) in Baja California, or 75 mm per degree of latitude. In Chile precipitation increases from about 350 mm at Valparaiso ($33°S$) to 2,489 mm at Valdivia

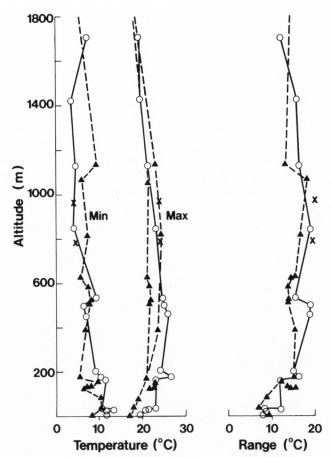

FIGURE 3-7. *Annual mean maximum and minimum temperature and range (°C) at different altitudes in southern California and central Chile. In California (O) the line connects stations on the coastal side of the mountain, and X's are stations on the inland side. In Chile (▲) the four stations at 140 m are located close to each other, as are the four stations at 500-625 m. These stations appear dominated by local microclimatic variations. Two lines are drawn; the cooler stations are in the vicinity of Santiago, the warmer stations are in the valley of the Rio Aconcagua.*

(39°48'S), 315 mm per degree of latitude, and decreases to about 2 mm at Iquique (20°24'S) in the Atacama Desert, or 28 mm per degree of latitude (Fuenzalida Ponce, 1967; Schneider, 1969). The steeper gradients toward the pole in Chile may lead to the smaller amplitude of the variations in annual precipitation in California than in Chile (Figure 3-6). Temperatures of the two countries are also similar in the mediterranean region. In both countries

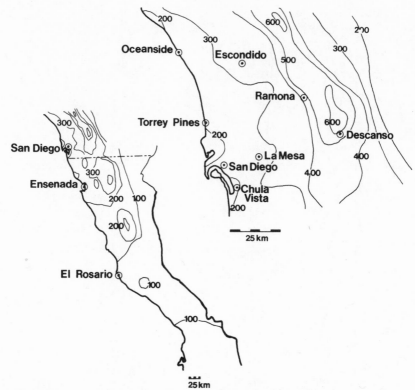

FIGURE 3-8. *Isopleths of annual precipitation (mm) for southern California, northern Baja California, (above) and central Chile (at right).*

mean minimum temperatures of the coldest month are 6-10° C, and mean maximum temperatures are 20-24° C.

Within the framework of broad climatic similarities, several general climatic differences occur. Tropical storms form during the fall in each hemisphere. In the Northern Hemisphere the storms develop at about 15° latitude and migrate northeastward, occasionally to northern Baja California and southern California. Thus, southward from latitude 30°30'N (San Quintin in Baja California) to about latitude 27°N precipitation gradually shifts from predominantly winter to predominantly late summer, although the annual total remains similar at about 100 mm (Hastings, 1964). At the latitude of San Diego these storms occasionally yield precipitation in the summer and early fall. Additionally, moisture from the Gulf of Mexico penetrates to southern California, and combined with orographic and convectional lifting of the air in the mountains, increases the precipitation in July and August in the mountains. In the Southern Hemisphere these tropical storms develop at about 5° latitude and rarely migrate as far south as northern Chile, much less central Chile. In Chile annual precipitation decreases toward the equator

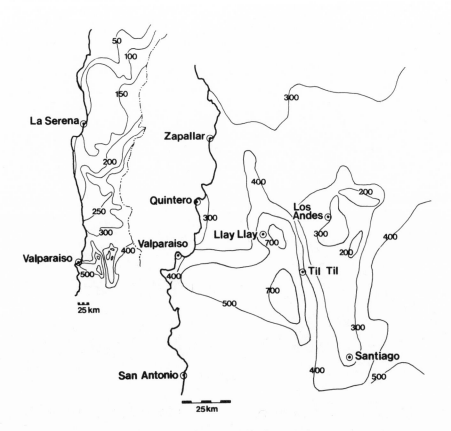

without a seasonal shift (Fuenzalida Ponce, 1967; Schneider, 1969). The Atacama Desert, lying south of the summer rains and north of the winter rains, is one of the driest regions on earth. Precipitation increases later in the fall and continues one month later in the sping in southern California than in central Chile (Figure 3-10), which affects the seasonal progressions of temperatures in the two countries (di Castri, 1973). Thus, the seasonal progression and latitudinal gradients of precipitation differ in the two countries.

On a global basis annual temperatures are lower in the Southern Hemisphere than in the Northern Hemisphere at similar latitudes (Sellers, 1965). These differences are reflected in the temperature of California and Chile at similar latitude (see pp. 62-64). Temperatures along the coast are modified by the ocean currents. Mean temperatures increase slightly toward the equator in southern California and Baja California as the cold California Current turns to the west, away from the coast; in Chile the Humboldt Current flows north close to the coast and temperatures increase less toward the equator. However, local variations in temperature caused by variations in topography or ocean currents can mask the regional trends. For example, mean monthly temper-

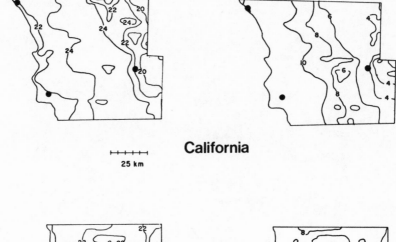

California

25 km

Chile

25 km

Mean maximum temp.   ## Mean minimum temp.

FIGURE 3-9. *Isotherms of mean maximum temperature and mean minimum temperature (°C) in southern California and central Chile.*

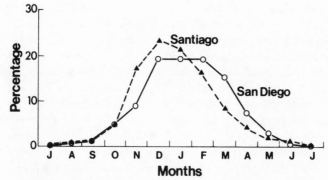

FIGURE 3-10. *Percentage of annual precipitation (mm) occurring in each month at San Diego, California, and Santiago, Chile.*

atures at Camp Pendleton and Oceanside are about 2° C higher than at Chula Vista, which is 50 km to the south.

In southern California high atmospheric pressure can develop east of the mountains and move dry air (the Santa Anas) from the desert toward the coast. This air warms adiabatically while descending from the mountains and can create hot, dry conditions of two to four days duration, usually in the fall (September and October). These warm air masses affect the mean, mean maximum, and absolute maximum temperatures (Figure 3-11), augment a continental aspect in the climate, and may create periods of high thermal and moisture stress for organisms. These desert air masses sometimes do not descend to the ground along the coast, and therefore create conditions of strong inversion, radiative cooling, and heavy fog and dew at night. In the Southern Hemisphere air movements from the deserts of Argentina toward Chile are blocked by the Andes, allowing more moderate and maritime conditions to prevail in Chile.

A fourth general climatic difference between the countries is that because the earth is closer to the sun on December 21 than on June 21, the Northern Hemisphere receives less solar radiation above the atmosphere during its summer than the Southern Hemisphere does during its summer at the same latitudes (List, 1963). Conversely the Northern Hemisphere receives slightly more solar radiation at the top of the atmosphere during its winter than does the Southern Hemisphere during its winter. At the 32°30' latitude this difference in the summers amounts to about 60 cal cm^{-2} day^{-1} more solar radiation above the atmosphere in the Northern Hemisphere on June 21 than in the Southern Hemisphere on December 21; and in the winter amounts to about 50 cal cm^{-2} day^{-1} more solar radiation in the Northern Hemisphere on December 21 than in the Southern Hemisphere on June 21. If the atmospheric transmission is the same in both hemispheres in the same season, the solar radiation at the earth's surface should be less in the summer in southern California than in Chile and more in the winter in southern California than in

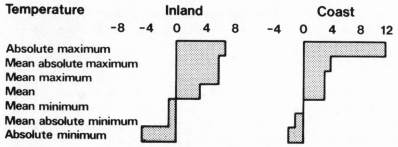

FIGURE 3-11. *Comparison of annual temperatures (°C) in California and Chile along the coast and inland. Values are the differences of Chile from California.*

Chile. On the average, cloudiness is greater in the Southern Hemisphere than in the Northern Hemisphere at similar latitudes (Sellers, 1965).

Regional Climatic Variations

A regional picture of the climate near 32°30' latitude can be gained from the records of precipitation and temperature that are available for the past twenty years from established weather stations in California and Chile. Unfortunately the stations are concentrated along the coast in California and in major river valleys in Chile. Thus, more than half the stations in each country are located below 250 m elevation, beneath the inversion in the maritime air. The temperature stations are better distributed throughout southern California and northern Baja California than in Chile. Temperatures at higher elevations, particularly in Chile, are approximate because of the lack of temperature-recording stations at higher elevations above the inversion.

The penetration of marine air and its moderating effect on daytime temperatures also follows the river valleys, but without orographic lifting of the air, and precipitation is much reduced. The precipitation along the broad, straight valleys in Chile does not change from the coast inland. The valleys in California are more irregular and rise more steadily than those in Chile, which are relatively broad, straight, and flat until the Andes are reached. Thus, isopleths of precipitation tend to parallel the coast in California but in Chile are often perpendicular to the coast (Figure 3-8).

The regional patterns of annual precipitation are similar in the two countries, and correlate with similar patterns of vegetation. In southern California, the annual precipitation along the coast at about 32°30' latitude is about 250 mm per year. At about 1,000 m, where evergreen shrubs predominate, the precipitation is about 550 mm per year, and in the mountains at 2,000 m, where deciduous and evergreen trees occur, it is about 650 mm per year. In Chile the precipitation along the coast at the same latitude is about 350 mm per year. The matorral appears to coincide with an annual precipitation of about 550 mm per year, and the *Nothofagus* forest in the coastal range with about 750 mm of precipitation per year. The succulent zone on the inland side of the coastal mountains below the matorral receives about 350 mm of precipitation per year. At Los Vilos, where the succulent zone extends to the coast, the coastal precipitation is about 300 mm per year. The acacia savanna in the central valley coincides with about 250 mm of precipitation per year. Precipitation in the broad valley of the Rio Aconcagua remains low from the coast inland because of the low elevational rise until the Andes are reached.

In contrast to arid regions with predominantly summer precipitation, where the regional pattern of precipitation is irregular and random (Noy-Meir, 1973), precipitation in the Chilean and Californian mediterranean scrub regions occurs in widespread storms. MacDonald (1956) distinguished between the widespread precipitation received from the winter storms that pass through southern California and Arizona and the localized precipitation received from convectional storms during the summer in Arizona. In both California and Chile correlations of annual precipitation between different stations in each country are usually high (Table 3–2). For example, the correlation of annual precipitation between San Diego and Descanso is 0.90, and between San Diego and Alpine, 0.92. The correlations between San Diego and Morena and between San Diego and Pine Valley are lower, because Morena and Pine Valley are in the mountains and receive summer rains and are in the rain shadow of coastal-facing mountains. In Chile the correlation of annual precipitation between Santiago and Valparaiso is 0.92, and between Santiago and Tiltil, 0.95. The regional patterns of precipitation affect the regional patterns of animal migration and activity, and the sporadic distribution of precipitation in many regions leads to a predominance of migratory animals (Noy-Meir, 1973).

The temperature records also indicate the effect on local climatic patterns of the rolling upland topography and of the long river valleys extending from mountains to coast. In California the mountains are evidently low enough and the river valleys circuitous enough that cold air flowing from the mountains does not greatly affect valley weather records. But in Chile stations in the

TABLE 3–2 *Linear Correlation Coefficients, and Regression Coefficients Relating the Annual Precipitation of an Established Precipitation Station to Precipitation at San Diego or Santiago*

Station	n^*	m	b	r
California				
Oceanside	20	0.94	40.0	0.93
San Vicente (B.C.)	9	0.78	27.8	0.79
Ensenada (B.C.)	45	0.76	93.0	0.60
Alpine	16	1.16	116.7	0.92
Descanso	19	1.79	169.5	0.90
Julian	21	1.54	279.9	0.85
Pine Valley	15	1.23	256.7	0.67
Morena	43	1.08	208.3	0.62
Chile				
Valparaiso	45	1.25	31.3	0.92
Tiltil	14	0.98	20.8	0.95
La Ligua	37	1.32	–90.7	0.88
Zapallar	10	0.97	73.9	0.68

$^*n$ is the number of years of overlapping record, m is the slope, b is the intercept, and r is the correlation coefficient.

long, broad valleys extending from the Andes to the coast, such as the valley of the Rio Aconcagua, in which Llay Llay and Los Andes are situated, and of the Rio Maipo near Santiago, have lower minimum temperatures and more frequent subzero temperatures than stations in the surrounding upland areas (Figure 3-9).

In both countries the geographic distribution of temperatures indicates the influence of regionally stable air and of topography; and in California the distribution also indicates the effects of the desert air masses (the Santa Anas). Mean annual temperatures along the coast and inland are about 2° C higher in southern California than in central Chile (Figure 3-11). In both countries the mean annual temperatures along the coast are similar to those of inland stations at the same elevation. The maximum annual temperatures show the effect of the Santa Anas and the more continental conditions in California. The mean maximum temperatures along the coast are about 3° C higher and inland about 6° C higher in California than in Chile. The absolute maximum temperatures are especially affected by the Santa Anas with absolute maximum temperatures 12° C higher along the coast and 7° C higher inland in California. Mean minimum annual temperatures along the coast are the same in California as in Chile, but inland the mean minimum temperatures are about 1° C lower in California than in Chile. Absolute minimum temperatures are 2° C lower at the coast and 5° C lower inland in California. Subzero temperatures have occurred at every station in California, and at the inland stations in Chile; subzero temperatures are uncommon at the coast in Chile.

In southern California minimum temperatures always occur in January and maximum temperatures occur in July in the mountains and in August at the lower stations. The maximum temperatures seem to be dominated by the radiation regime in the mountains and by the abiabatic warming of the air during the Santa Ana winds in late summer at the lower elevations. In the mountains temperatures below freezing can be expected from October to May, but are always above freezing from June to September. In Chile minimum temperatures always occur in July, and maximum temperatures always occur in January. No temperatures are available for the coastal mountains in Chile.

In both countries diurnal and seasonal variation in air temperature increases from the coast to the interior, but more so in southern California. Los Andes, among the Chilean stations, is the most similar to the inland station in southern California in diurnal and seasonal variation in air temperature.

PHYSICAL ENVIRONMENT OF THE PRIMARY SITES

The physical environment affects the physiology and behavior of the organism through the physical processes of energy exchange, i.e., through radiation, convection, evaporation, and conduction (Gates, 1962, 1965). The first response of the individual organism to variations in the processes of energy exchange is a change in its temperature, or changes in behavioral or physiological responses related to temperature control. Later responses of the individual are those relating to physiological or behavior responses to temperature, possibly acclimation or genetic changes in the population. The processes of energy exchange can be compared at both a regional and a local level. Budyko (1956) analyzed the seasonal patterns of the partitioning of net radiation into convection and evaporation. In contrast to arid deserts in which convection is a large fraction of net radiation throughout the year or to rain forests in which evaporation is a large fraction of net radiation throughout the year, the mediterranean scrub region shows high convectional exchange during the summer and high evaporational exchange during the winter relative to the net radiation received in these periods.

The climates of the primary sites were measured so that the processes of energy exchange could be estimated, in relation to their effect on individual organisms and vegetation. In arid and semiarid environments the primary control of ecosystem processes is related to the availability of water, so the semiarid environment should be conceived first in water terms and secondarily in energy terms (Noy-Meir, 1973). Thus, in this section the water environment and evaporation potential will be discussed first and then the other processes of energy exchange.

Water

Precipitation along the coast in both California and Chile is relatively light and occurs in storms of low average rainfall per storm. Proceeding inland into the higher elevations, total precipitation, rainfall per storm, and number of storms increase. During the period of the project, in both countries, precipitation occurred predominantly during the winter, and precipitation was higher at inland sites than at coastal sites. Precipitation varied during the period of the project (Figure 3-11). In California the 1972-73 water year was a wet year and the others were relatively dry. In Chile 1972 was a wet year and 1973 was more normal. The seasonal distribution of precipitation differed slightly between the two countries: the mountains of California receive some summer precipitation, and some light precipitation occurred during the summer at the California primary site. In Chile no precipitation occurred during the summer at any of the research sites.

Not all incoming precipitation is available for plant or animal consumption. Some of the incoming precipitation is intercepted by the vegetation canopy and evaporated, some is evaporated from the soil surface, and some is lost to plants because of surface and subsurface runoff although this water may become available to animals from ponds and streams. The amount of available water depends on the intensity of frequency of storms. At the San Dimas Experimental Station in southern California (Hamilton and Rowe, 1949; Rowe and Colman, 1951) the annual rainfall intercepted by the vegetation canopy and evaporated was about 8-11 percent of the incoming precipitation. This percentage increases with light rains and decreases with heavier rains, because the canopy has a water storage capacity that must be reached before significant amounts of intercepted water drip from the canopy or run down the stems. This water-storage capacity is probably in the range of 0.16–0.76 mm per unit of leaf area index (one side), a range including various grasses, shrubs, and trees (Leyton et al., 1967).

Water reaching the soil surface may run off the surface or infiltrate into the soil. Water that infiltrates the soil and is held in the upper 0.05 to 0.10 m is readily lost by evaporation and scarcely available for plant use. Root systems excavated in Californian chaparral by Hellmers et al. (1955) and in both the Californian chaparral and Chilean matorral by Ng (unpublished data) indicate several species with roots predominantly above 0.30 m depth and others with deeply penetrating taproots. Although these depths do not sharply divide actual root or soil zones, the discussions that follow will treat these zones as discrete units for simplicity. Water penetrating to about the 0.10–0.30 m depth will be considered available for shallow-rooted annuals, perennials, and shrubs, and water penetrating below 0.30 m will be considered available for deeper-rooted shrubs.

The depth to which water will infiltrate depends upon the total precipitation, number and frequency of storms, precipitation received in individual storms, and soil composition. The maximum quantity of water that can be held in the soil is about 0.43 cm^3 per cm^3 of soil, in both countries; the rest of the soil volume consists of mineral constituents. About 0.25 cm^3 of water per cm^3 of soil can be held against the force of gravity. Water in excess of this amount passes readily into the deeper soil. Water in lesser amounts flows through the soil across gradients of water potential and hydraulic conductivity. The driest soils measured in the field contained about 0.05 cm^3 of water per cm^3 of soil, and the moisture profiles indicated even drier soils at the surface where evaporation was occurring. Thus, about 20 mm of precipitation (0.20 cm^3 cm^{-3} through a depth of 0.10 m) can be held in the surface layers of the soil and readily lost by evaporation. Moisture from light rains during the summer and early fall, when the soil is dry, is held in the surface layer of soil, usually evaporating before the next storm, and is relatively unavailable for plant use. During the fall and winter the amount of water evaporated between storms must be replenished in the soil before significant amounts infiltrate to the root zone of shallow-rooted plants. About 60 mm of precipitation is required for significant amounts of water to infiltrate to the root zone of the deeper-rooted shrubs.

When the precipitation occurs in separate storms, so that evaporation can occur between storms, the minimum amounts of precipitation required for infiltration are increased. For example, if all the water of a storm is held in the upper layer of soil and evaporated before the next storm, and if the incoming precipitation is divided equally among eight storms, approximately 240 mm of annual precipitation would be required for water to infiltrate below the shallow-root zone; but if the precipitation is divided among ten storms, about 260 mm would be required for deep infiltration. The amount of precipitation required for water to infiltrate into deeper levels is decreased if the evaporation between storms is reduced, as occurs during the winter, or if some storms contribute more water than others, or if the assumption of a water-storage capacity is altered to include water flow along a water-potential gradient. However, it should be kept in mind that intense storms will have greater surface runoff.

In broad form this concept seems applicable to mediterranean scrub regions. In both California and Chile shallow-rooted shrubs predominate with an annual precipitation averaging 250 mm, which accumulates during approximately eight storms. Deep-rooted shrubs predominate with an annual precipitation of about 550 mm, which accumulates during about ten storms. Shachori and Michaeli (1965), after reviewing twenty-six studies of water yield in regions with less than 1,400 mm annual precipitation, suggested that 400 mm year^{-1} was the minimum precipitation required for the development of woody vegetation (forest, woodland, or maqui).

Available soil moisture also varies with soil texture. Finer-textured soils hold more water per unit volume than do coarser-textured soils. With a region of similar precipitation, finer-textured soils will hold more water closer to the surface where it is readily lost by evaporation and be drier than coarser-textured soils. Thus, finer-textured soils may be populated more abundantly by shallow-rooted species such as herbs and forbs, although the amount of annual precipitation indicates that deeper-rooted plants could occur. The soils on the ridgetop at the primary site in Chile are finer-textured than the soils on the slopes in Chile or at the California primary site, and thus should support relatively larger populations of herbs and forbs.

Evaporational and transpirational losses of soil moisture relate to all the processes of energy exchange. The rate of water loss under an existing incoming energy regime, if the resistances of the soil and plants to water loss are minimal, is expressed by the potential evapotranspiration rate, which can be estimated according to Thornthwaite (1948) and Penman (1948). The potential evaporation is minimal in winter and maximal in summer in both California and Chile (Table 3-3); varying between 2.3 and 9.8 mm day^{-1} in California and 2.8 and 7.8 mm day^{-1} in Chile using the Thornthwaite method, and between 0.2 and 8.2 mm day^{-1} in California and 1.8 and 7.6 mm day^{-1} in Chile using the Penman method (Table 3-3 and 3-4).

The seasonal pattern of soil-moisture content is the result of the seasonal patterns of precipitation and evapotranspiration. In California and Chile, during the period of the project, soil moisture increased through the winter and decreased through the summer (Table 3-3; Figure 3-12). At the coast in California the relatively small amount of precipitation in each storm resulted in most of the moisture being stored in the surface layer of soil. Incoming precipitation often did not appear at the 0.30 m depth. In both countries,

TABLE 3-3 *Average Monthly Climatic Indices at the Primary and Coastal Sites in California and Chile, April 1971–December 1974*

	Jul.	Aug.	Sep.	Oct.	Nov.	Dec.	Jan.	Feb.	Mar.	Apr.	May	Jun.
California Primary												
Solar radiation (cal cm^{-2} day^{-1})	558	529	453	330	265	220	249	317	288	502	528	571
Daylength (h min)	14 16	13 25	12 21	11 16	10 18	09 48	10 06	11 01	12 01	13 08	14 03	14 30
Net radiation (cal cm^{-2} day^{-1})	386	399	290	230	143	89	118	190	189	352	420	371
Precipitation (mm mo^{-1})	7.6	7.7	4.2	50.3	62.8	47.2	81.5	45.8	78.9	39.6	12.8	10.6
Potential evapotranspiration (mm day^{-1})	6.6	6.5	5.1	3.8	3.0	1.8	1.6	2.7	2.8	4.3	5.2	5.2
Soil moisture, 0.3 m to bedrock (mm)	92	87	85	90	92	100	116	136	124	148	185	110
Air temperature max/min (°C)	32 / 11	31 / 11	28 / 9	22 / 7	18 / 3	15 / 2	15 / 1	17 / 2	19 / 6	19 / 2	22 / 6	28 / 8
Ridge soil temperature at 2 cm max/min (°C)	37 / 20	37 / 18	35 / 15	27 / 10	20 / 8	7 / 6	14 / 6	17 / 7	14 / 7	21 / 8	32 / 17	32 / 17
Ridge soil temperature at 32 cm max/min (°C)	28 / 23	28 / 23	24 / 22	18 / 16	14 / 13	12 / 10	10 / 9	11 / 10	11 / 10	14 / 11	25 / 16	23 / 19
Midday relative humidity (%)	36	43	46	53	57	58	58	42	43	40	44	38
Absolute humidity max/min (g m^{-3})	14 / 9	15 / 9	14 / 8	10 / 7	9 / 5	7 / 5	8 / 5	6 / 5	8 / 5	7 / 5	9 / 7	10 / 7

California Coastal

Measurement												
Solar radiation (cal cm⁻² day⁻¹) — $\text{cal cm}^{-2}\,\text{day}^{-1}$	459	428	374	295	242	198	212	211	327	455	402	451
Daylength (h min)	14 16	13 25	12 21	11 16	10 18	09 48	10 06	11 01	12 01	13 08	14 03	14 30
Net radiation ($\text{cal cm}^{-2}\,\text{day}^{-1}$)	322	291	231	247	98	166	124	67	152	332	317	305
Precipitation (mm mo^{-1})	0.3	0	2.9	9.4	41.9	34.8	22.7	29.9	43.6	5.9	5.6	6.0
Potential evapotranspiration (mm day^{-1})	4.9	4.6	3.7	4.0	1.7	2.5	1.8	1.5	2.2	4.5	4.3	4.5
Soil moisture, 0.3 m to bedrock (mm)	160	158	164	170	163	178	177	180	181	182	172	160
Air temperature max/min (°C)	23 / 16	24 / 17	23 / 17	21 / 14	17 / 12	17 / 11	15 / 10	17 / 11	16 / 11	19 / 11	19 / 13	22 / 15
Midday relative humidity (%)	63	62	61	58	60	56	60	66	70	56	65	62
Absolute humidity max/min (g m^{-3})	14 / 12	14 / 13	13 / 12	10 / 9	9 / 8	8 / 6	8 / 7	9 / 7	10 / 9	9 / 8	12 / 10	13 / 11

TABLE 3-3 *(Continued)*

	Jan.	Feb.	Mar.	Apr.	May	Jun.	Jul.	Aug.	Sep.	Oct.	Nov.	Dec.
						Chile Primary						
Solar radiation (cal cm^{-2} day^{-1})	611	539	457	334	196	150	193	264	360	489	575	635
Daylength (h min)	16	25	21	16	18	48	66	01	01	08	03	30
	14	13	12	11	10	09	10	11	12	13	14	14
Net radiation (cal cm^{-2} day^{-1})	452	426	367	236	151	126	140	307	289	383	501	501
Precipitation (mm mo^{-1})	0	0	0	1.6	14.2	728.5	99.2	95.2	48.7	30.7	1.3	0
Potential evapotranspiration (mm day^{-1})	6.9	6.7	5.9	4.1	2.3	2.1	2.3	3.7	3.8	5.2	6.7	7.3
Soil moisture, 0.3 m to bedrock (mm)	110	136	162	114	146	138	137	154	197	191	163	204
Air temperature max/min (°C)	23	23	22	20	15	11	12	13	14	17	19	21
	11	10	9	8	7	5	4	5	5	7	8	10
Ridge soil temperature at 2 cm max/min (°C)	43	37	32	27	19	12	13	19	25	22	36	46
	17	15	13	13	9	6	5	4	9	10	13	14
Ridge soil temperature at 32 cm max/min (°C)	23	23	20	18	14	11	9	9	13	13	19	20
	22	22	19	18	14	10	9	8	12	12	18	19
Midday relative humidity (%)	45	45	41	40	54	64	55	55	53	54	47	47
Absolute humidity max/min (g m^{-3})	10	9	8	7	7	7	6	6	7	8	8	9
	9	9	8	6	6	5	4	5	5	7	7	8

Chile Coastal

Solar radiation (cal cm^{-2} day^{-1})	494	368	303	196	158	127	157	205	260	339	409	465
Daylength (h min)	16 14	25 13	21 12	16 11	18 10	48 09	06 10	01 11	01 12	08 13	03 14	30 14
Net radiation (cal cm^{-2} day^{-1})	351	243	134	76	87	68	72	177	180	245	363	372
Precipitation (mm mo^{-1})	0	0.2	0.4	0.2	50.7	152.5	38.7	0	18.4	28.0	2.6	0.2
Potential evapotranspiration (mm day^{-1})	4.8	3.2	2.4	1.1	1.1	1.0	1.0	2.3	2.1	3.0	4.4	4.6
Soil moisture, 0.3 m to bedrock (mm)	–	–	–	–	–	–	–	–	–	–	–	–
Air temperature max/min (°C)	21 14	20 14	18 13	16 12	15 13	14 11	13 10	14 9	14 10	16 11	18 13	20 14
Midday relative humidity (%)	83	89	88	93	92	94	94	90	86	84	79	81
Absolute humidity max/min (g m^{-3})	14 12	15 12	14 12	12 10	12 11	12 10	11 9	10 9	10 9	11 9	12 11	14 12

TABLE 3-4 *Mean Daily Total Solar Radiation, Net Radiation, Potential Evapotranspiration Calculated from Penman (1948) and Thornthwaite (1948) Relations, and Soil Moisture Withdrawal for the 0.3–1.2 m Depth on Equator-facing, Pole-facing, and Ridgetop Surfaces at the Primary Sites.*

| | California | | | | | | | | | | | |
	July	Aug.	Sep.	Oct.	Nov.	Dec.	Jan.	Feb.	Mar.	Apr.	May	June
Solar total radiation (cal cm^{-2} day^{-1})												
Equator-facing slope	610	623	606	560	385	363	321	291	252	588	609	586
Ridgetop	649	640	513	517	333	299	270	283	246	561	610	616
Pole-facing slope	619	578	492	378	219	177	355	260	230	486	550	580
Net radiation (cal cm^{-2} day^{-1})												
Equator-facing slope	548	553	532	490	341	322	186	202	203	391	424	348
Ridgetop	386	399	290	230	143	89	118	190	189	352	420	371
Pole-facing slope	328	310	165	70	19	-42	22	181	189	266	345	310
Potential evapotranspiration (Penman) (mm day^{-1})												
Equator-facing slope	8.2	8.2	7.6	6.4	4.2	3.8	2.4	2.8	2.9	4.8	5.5	5.3
Ridgetop	6.1	6.3	4.6	3.4	2.1	1.5	1.7	2.7	2.8	4.5	5.5	5.6
Pole-facing slope	5.4	5.1	3.0	1.6	2.6	0.2	0.8	2.6	2.8	3.6	4.6	4.8
Potential evapotranspiration (Thornthwaite) (mm day^{-1})												
Ridgetop	9.8	9.6	7.5	5.0	3.0	2.3	2.4	3.1	3.6	4.0	5.3	7.6
Soil moisture withdrawal (mm day^{-1})												
Ridgetop	0.7	0.4	0.1	0	2.0	0.5	2.4	1.3	1.7	1.4	0.5	1.8
Pole-facing slope	0.9	0.2	0	0	3.0	0	1.5	1.2	2.0	1.5	1.0	2.1

Chile

	Jan.	Feb.	Mar.	Apr.	May	June	July	Aug.	Sep.	Oct.	Nov.	Dec.
Solar total radiation (cal cm^{-2} day^{-1})												
Equator-facing slope	678	658	613	456	328	270	288	517	472	556	688	684
Ridgetop	720	670	578	414	286	230	249	446	451	563	737	756
Pole-facing slope	678	593	449	296	188	143	161	273	361	502	702	720
Net radiation (cal cm^{-2} day^{-1})												
Equator-facing slope	453	440	413	269	143	102	177	355	316	397	484	471
Ridgetop	452	426	367	236	151	126	140	307	289	383	501	501
Pole-facing slope	435	385	281	153	79	60	68	166	229	339	485	525
Potential evapotranspiration (Penman) (mm day^{-1})												
Equator-facing slope	6.9	7.0	6.6	4.6	2.5	2.2	2.8	4.5	4.1	5.5	6.6	7.0
Ridgetop	6.9	6.8	6.1	4.3	2.5	2.4	2.4	4.0	3.9	5.3	6.8	7.3
Pole-facing slope	6.7	6.3	5.1	3.3	1.8	1.8	1.8	2.7	3.3	4.8	6.7	7.6
Potential evapotranspiration (Thornthwaite) (mm day^{-1})												
Ridgetop	7.4	7.8	7.4	6.5	5.1	3.5	2.8	2.6	3.2	4.3	6.5	7.4
Soil moisture withdrawal (mm day^{-1})												
Ridgetop	0.9	0	0.8	1.2	–	5.4	5.4	–	3.1	0.6	0.7	0.7
Pole-facing slope	–	–	–	–	4.8	4.8	4.8	4.8	2.9	0.6	1.0	–

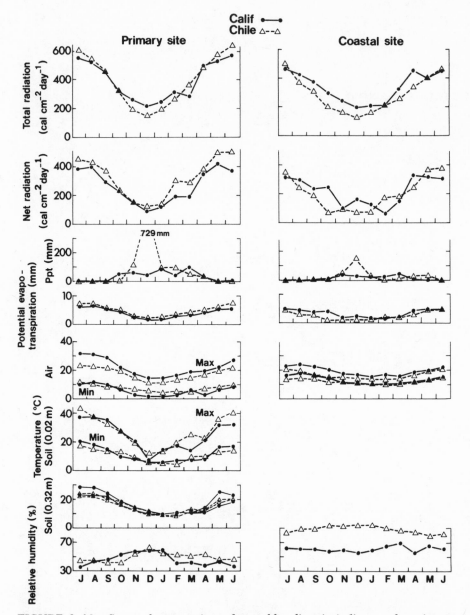

FIGURE 3-12. *Seasonal progression of monthly climatic indices at the primary coastal sites in California and Chile. Values are averages for the period April 1971 to December 1974 for climate.*

at both coastal and inland sites, soil moisture decreased to minimum levels well before the next rainy season, although the amount of moisture held in the soil at this minimum level varied by site and slope. In California soil moisture reached minimum levels earlier at the coast than at the inland site. Soil moisture was minimal from early July through October at the coast and from mid-August through October inland. In Chile soil moisture appears to be high from the winter rain through November. Minimum levels are reached in January.

Throughout the year soil moisture was more abundant on the equator-facing slope than on the pole-facing slope at the inland sites in California and Chile and at the coastal site in California between 0.2 and 1.2 m depth. However, near the surface soil moisture was depleted about one month earlier on the equator-facing slope than on the pole-facing slope. Soil moisture in the oakgrove in California, located in a shallow valley, was greater throughout the year than in either the equator- or pole-facing slopes.

The moisture withdrawal from the soil, which should be related to potential and actual evapotranspiration, was maximal in winter and spring in both countries. During the winter moisture withdrawal accounted for most of the energy received in net radiation; in Chile moisture withdrawal exceeded net radiation. Soil moisture withdrawal in these months was about 2 mm day^{-1} in California and 5 mm day^{-1} in Chile (Table 3-4). The excess is probably due to surface and subsurface runoff. During the spring evapotranspiration became a decreasing fraction of the net radiation, and during the summer moisture withdrawal was negligible for about 2 1/2 months. Net radiative energy that is not used in evaporation will largely be lost by convection, since conduction into the soil is small over weekly or monthly periods. Moisture withdrawal at all depths is greater from the dense vegetation on the pole-facing slopes than from the sparse vegetation on the equator-facing slopes and ridgetop.

The availability of water for transpiration and plant growth depends upon the water potential of the soil, rather than upon the absolute amount of water in the soil. The water potential, which is an expression of the energy state and availability to plants of water in the soil, was derived from moisture-tension curves for the soils of each country (Ng, 1974; Poole, unpubl. data). The moisture tension curves for California showed water potentials above –2 bars with moisture contents greater than 0.15cm^3 water cm^{-3} of soil, and water potentials decreasing abruptly with moisture contents below 0.15 cm^3 cm^{-3}. The moisture tension curves for Chile showed abrupt decreases in water potential with water contents below about 0.09 cm^3 cm^{-3} on the pole- and equator-facing slopes, but the decrease occurs below about 0.23 cm^3 cm^{-3} on the ridgetop, due to the finer-textured soils on the ridgetop. Thus, in California during the summer, soil water potentials on the equator-facing slope were –7 to –10 bars at 0.30 m and –1 to –2 bars below 0.40 m depth. Water potentials in the root zone of the shallow-rooted plants were lower than in the root zone of the deep-rooted shrubs. In the winter, water potentials were high at all depths. On the pole-facing slope, soil water potentials at all depths were -7 to -13 bars during the summer and about -1 bar during the winter. The pole-facing slope at the coast also had lower potentials than the equator-facing slope, although potentials below 0.30 m were usually above -1 bar on both slopes. In Chile the soil water potentials in the deeper levels were similar at analogous times of the year to those in California.

Plant water potentials indicate the state of water in the plant, and when transpiration is negligible, such as just before dawn, can indicate the soil moisture potential in the vicinity of the roots. Plant water potentials during the summer were considerably lower than the soil water potentials, indicating low exploitation of the root zones by the shrubs and low hydraulic conductivity of the soil at low water potentials. Predawn plant water potentials in California were -30 to -50 bars by the end of the summer (Poole and Miller, 1975). In Chile comparable plant water potentials were -6 to -10 bars in the summer of 1972-73 (Giliberto, unpubl. data), and soil water potentials were similar. This summer followed an unusually wet winter (Figure 3-13).

In California the length of the summer drought, whether measured by soil moisture content, by water potential, or by the more functional measurements of plant water potential and leaf conductance, is one to two months longer at the coast than inland (Poole and Miller, 1975). The period of soil drought at the coast, as measured by these various methods, is more than ninety days, whereas at the primary site the soil drought is less than ninety days.

Absolute and Relative Humidity

The actual and potential evaporation rates depend in part upon the absolute and relative humidities of the air. The absolute humidity, measured on the hygrothermograph at the inland sites was generally higher in California than in Chile, especially during the day; but the relative humidity tended to be lower during the day and higher during the night in California than in Chile (Table 3-3). The differences in relative humdity reflect the more variable air temperatures and the more variable saturation water contents of the air in California. In both countries the absolute humidities varied by only 1-2 g m^{-3} through the day, and were near saturation humidities at night. In both countries the annual variation in absolute humidity was greater than the diurnal variation. At the coast the absolute and relative humidities were lower in California than in Chile; coastal temperatures are lower in Chile, thus decreasing the saturation vapor density of the air. The humidities measured may also have been influenced by the different topographic settings of the two sites. The California coastal sites are more exposed to offshore breezes and less subject to ocean spray than is the Chilean coastal site.

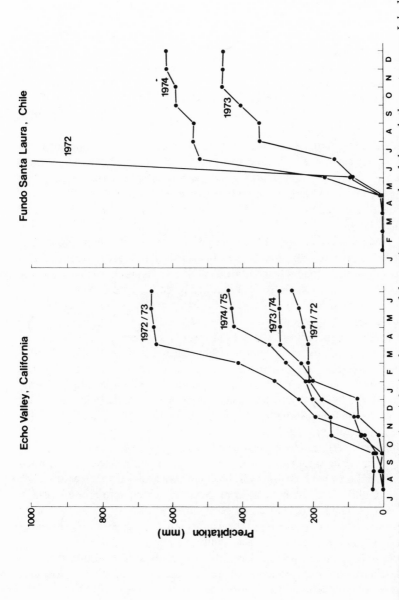

FIGURE 3-13. *Precipitation (mm) during the years of the project accumulated through the water year: July 1–June 30 at San Diego (above), January 1–December 31 at Santiago (at right). In 1972 Fundo Santa Laura received 2260 mm.*

59

Radiation

Radiation directly affects the temperature of the soil and organisms, photosynthesis, and indirectly rates of evaporation and transpiration and growth. Radiation is considered in terms of solar irradiance, which is predominantly in the short wavelengths (300–3,000 nm), and infrared irradiance, which is in the long wavelengths (9,000–11,000 nm). Solar irradiance was less in California than in Chile during the summer but greater in California than in Chile during the winter (Table 3–4). In both countries solar irradiance was about 100 cal cm^{-2} day^{-1} less at the coast than at the primary sites because of the increased cloudiness at the coast. In California solar irradiance in the mountains was similar to the solar irradiance at the coast. In both countries the highest irradiances measured were between 700 and 799 cal cm^{-2} day^{-1}. During the winter irradiances of less than 100 cal cm^{-2} day^{-1} were common in both countries and were very common in Chile during the unusually rainy and cloudy winter of 1972.

The solar irradiance incident on the vegetation or soil surface differs through the year on the different exposures because of the seasonal variation in incoming irradiance and the geometric relations of the incoming rays to the surfaces. The equator- and pole-facing slopes and ridgetop were carefully selected to ensure that the orientation of the comparable surfaces was the same in both countries. On the equator-facing slopes daily solar irradiance is greatest during April and again in late August to early September in California and during the analogous months in Chile (Frank and Lee, 1966). The daily irradiance in California decreases slightly between April and August, when the rays of the sun are more oblique to the surface. From early May through August in the Northern Hemisphere, and during the analogous months in the Southern Hemisphere, the irradiances on equator-facing slopes are similar to those on pole-facing slopes, because the lower midday intensities on the pole-facing slopes are compensated by the higher intensities during early morning and late afternoon.

Interception of the direct solar beam by the vegetation is almost complete on the pole-facing slope, and is greatest at the top of the canopy. The more open vegetation of the ridgetops in both countries and of the equator-facing slope in California allows more solar radiation to penetrate to the soil surface. The level at which interception of solar radiation occurs, i.e., at the top of the canopy or at the soil surface, affects not only the partitioning of energy by the soil and vegetation but also the profiles of air and soil temperature. Interception in the canopy increases the convectional loss of absorbed radiation, whereas the interception at the soil surface increases the infrared radiation loss and conduction into the ground.

Infrared radiation from the sky at the two primary sites varied through the course of the year between 0.50 cal cm^{-2} min^{-1} in the summer and 0.45 cal cm^{-2} min^{-1} in the winter. Low values of about 0.40 and high values of 0.55 cal cm^{-2} min^{-1} were measured in both countries.

The net radiation on the different slopes varied seasonally, owing in part to changes in the angle and therefore the intensity of the incoming solar irradiance, and in part to the different intensities of infrared radiation emitted by the vegetation and the soil surface (Table 3-4). The albedo, corrected for the orientation of the surface, increased slightly at low angles of incidence of the sun's rays to the surface. On ridgetop and pole-facing slopes, net radiation was lower in California than in Chile throughout the year, owing to the combined effect of solar irradiance and air temperature. Differences between slopes were greater in California, because of the differences in vegetation cover. In both countries almost all the net radiation is lost in evaporation through the winter from the surface. In midspring net radiation increases but evaporation stays constant, and convectional loss increases. Net radiation increases until early summer, but evaporation decreases through spring and summer because of the decreasing water supply. Through the summer most of the net radiation is lost by convection rather than evaporation. In the fall net radiation decreases, and by the time the rains begin, net radiation is low.

Wind

The processes by which water vapor and heat are lost from the vegetation/ soil surface or from an organism involve the turbulent exchange of air. This turbulent exchange is related to wind speed and to surface properties, such as the roughness of the topography and vegetated surface. Without turbulent exchange the air would become saturated with water near the surface, and air temperatures near the surface would rise.

Winds are predominantly from the west and northwest in both California and Chile; however, during the period of the project the wind speeds in California were about one-half those in Chile. Wind speeds also tended to be more erratic in California; in Chile, wind speeds tended to be fairly uniform through the year, although slightly higher in winter. In both countries the diurnal wind pattern at the primary sites consisted of low speeds at night, with light downslope winds, and higher speeds in the afternoon from the coast. At the Echo Valley primary site the changes in predominant wind direction occurred through the day. Wind was often from the east-northeast at night, shifting to the south for short periods during the day, and then to the northwest or north in the afternoon, bringing in air from the ocean; wind speeds increased when the winds shifted to the northwest. At the Fundo Santa Laura primary site westerly winds occurred throughout the day and night. The diurnal patterns of wind direction are related to topography at night and to the predominance of the land-sea breezes during the day. At the latitude of Fundo Santa Laura, wind is predominantly from the southwest. The predominance of northwest winds at the research site is believed to result from a relatively low ridge to the northwest, which allowed air to pass, while to the southwest and south, a high mountain ridge impeded air movement. At the coast wind speeds were higher than inland in California, whereas they were lower at the coast than inland in Chile. Wind speeds at the coast were lower in Chile than in California, a difference which probably reflects the topographic placement of the anemometers. In Chile the anemometer was located below the top of a coastal cliff, while in California it was on a coastal-facing ridge.

Air and Soil Temperatures

At the primary sites, diurnal variations in air temperature of $20°$ C were common in California, but variations were rarely greater than $10°$ C in Chile (Table 3-3). The higher wind speeds in Chile maintained the more moderate air temperatures. Air temperatures were $10°$-$30°$ C in California and $10°$-$20°$ C in Chile in the summer, and $1°$-$15°$ C in California and $4°$-$12°$ C in Chile in the winter. During the project, extreme temperatures (i.e., daily mean maximum temperatures above $30°$ C or mean minimum temperatures below $0°$ C) were more common at Echo Valley than at Fundo Santa Laura. Both the diurnal and the seasonal temperature variations were about twice as great at Echo Valley as at the California coast, the Mount Laguna (mountain) station, or Fundo Santa Laura. At the coastal stations in both countries, the diurnal variation of air temperature was about $5°$ C.

The turbulent exchange of air is greater during the day than at night, owing to the higher wind speeds during the day. The increased turbulent exchange tends to equalize maximum air temperatures on the different slopes. At each primary site weekly mean maximum air temperatures at the top of the vegetation canopy on the pole- and equator-facing slopes and ridgetop, as well as at the standard weather screens, were all within $1°$-$2°$ C of each other. Minimum temperatures occur at night when turbulent exchange is minimal; and these temperatures show greater variation at the primary sites than the maximum temperature. The mean minimum temperatures were $2°$ C lower in topographic depressions than on the slopes. Additional measurements indicate that air temperatures 0.10 m above the soil surface within or between shrubs, were within $1°$-$2°$ C of one another. Air temperatures through a vertical profile from the top of the canopy or shrub down to 0.10 m above the soil surface varied less than $1.0°$ C. Thus, the range of air temperature around the primary sites during the day was within $1°$-$2°$ C, and temperature measured at the weather screen could be used to estimate temperature around the sites with this level of accuracy. The relation between weekly mean maximum and minimum temperatures in the weather screen and on the slopes was summarized in linear regression equations.

Air temperatures affect, and are affected by, organism temperatures through the convectional exchange of energy. In both countries the highest air temperatures occured at the top of the canopy on the pole-facing slope, because radiation was intercepted at the top of the canopy. Simulation of the processes of energy exchange through the chaparral canopy indicate that the leaf temperatures are close to and often slightly below air temperature. Stem temperatures by contrast are about 1° C above air temperatures, because of their lower rate of transpiration. These temperature differences mean that about 80 percent of the convectional loss of energy from the canopy is from stems. The stems intercept solar irradiance and convect this energy away. Without the stem interception, leaves would intercept more solar irradiance and evaporative demands would be greater. If the higher interception irradiance caused no increase in photosynthesis (i.e., irradiances were above light saturation), the higher loss of water would be wasteful and cause decreased annual production.

The California sites seem to match the Chilean sites more closely in soil surface temperatures on the ridgetops than in air temperatures or solar irradiances. During the summer, the lower irradiances and higher air temperatures in California apparently compensate for each other, thus, where similar vegetation cover occurs in the two countries, similar soil surface temperatures also occur. However, temperatures at the 0.32 m depth on the ridgetops were higher and more variable through the day and through the year in California than in Chile.

Soil temperatures on the various slopes differed more at the California primary site than at the Chilean primary site, partly because the two countries differ in vegetation cover at locations where the soil temperatures were measured. The soil was completely covered by vegetation on the pole-facing slopes of both countries and on the equator-facing slope in Chile, but only about three-fourths covered on the equator-facing slope in California and ridgetops in both countries. The highest soil temperatures occurred at the surface of the soil under an open vegetation canopy. Soil temperatures over 45° C, which are detrimental for many organisms, were common from the surface to 0.10 m depth in spring and summer on the ridgetops in both California and Chile, and on the equator-facing slope in California. Although temperatures were more moderate in the plant root zones, these temperatures were higher under open vegetation than closed vegetation. In contrast to the similarity in air temperatures on the different slopes at our primary site, soil surface temperatures on different slopes varied widely. Temperatures on the pole-facing slopes were higher in the summer, but more variable during the day and through the year in California than in Chile. Weekly mean maximum and minimum soil surface temperatures correlated with the weekly mean maximum and minimum air temperatures measured in the weather screen, although the maxima and minima of air and soil temperatures all occurred at different times of day. Temperatures below 0° C occurred at the soil in both countries.

Net radiation is exchanged by convection, evapotranspiration, and heat conduction into the ground. Heat conduction into the ground varied widely through the day on exposed soils but varied less widely under the vegetation.

The heat conduction was between 0.10 and -0.05 cal cm^{-2} min^{-1} through the day and night. However, on a daily basis, heat conduction is a small fraction of the net radiation, amounting to less than 2 cal cm^{-2} day^{-1} throughout the season.

Thus, the mediterranean scrub vegetation of California experiences higher temperatures and slightly lower potential evapotranspiration than that of Chile. The winter rain-summer drought pattern of precipitation characteristic of both countries results in high evaporation relative to net radiation during the winter and higher convection relative to net radiation during the summer. When temperatures become optimal for photosynthesis and growth, water is consistently present in the soil. Early in the spring, temperatures at the soil surface are warmer and more optimal for physiological processes than temperatures in the air. Water is available for both shallow- and deep-rooted plants. At this time the grasses and forbs are active and the shrubs are beginning to be active. Late in the spring and during the summer, water is no longer available in the upper soil layers and soil surface temperatures are unfavorably hot in open canopies. However, water is available at depth and is tapped by the deep-rooted plants, which by bearing their leaves off the ground surface enhance convectional cooling and reduce leaf temperatures. The convectional loss of energy is further enhanced when the vegetation consists of scattered shrubs with small leaves. The placement of leaves in the air rather than at the soil surface, the small leaf sizes, and the open canopy structure all contribute to lower leaf temperatures and reduced evaporational demands. The high proportion of stem area to leaf area in both scattered shrubs and continuous canopies further reduces the evaporative demands on the leaves through the interception and convection of solar irradiance by the stems. Thus, many of the structural similarities of the vegetation in the two countries relate to the similarities in the physical environments, which affects the water balance, and ultimately, carbon balance of the plants.

GEOLOGIC HISTORY, LITHOLOGY, AND SOILS

The vegetation is shaped by climate and soils, and the soils, in turn, are shaped by climate, vegetation, and underlying rocks. The underlying rock, by influencing geomorphic processes which form the topographic exposure and diversity, affects the radiation, thermal, and water balance of the vegetation and, by influencing the soil texture and chemistry, affects the inorganic nutrition and water balance of the vegetation. The influences of topography on microclimate and the influences of rock composition on soil texture and soil chemistry are more pronounced in arid and semiarid regions than in humid regions. The affect of topography on microclimate is greater because of the unavailability of water for the temperature moderating influences of evaporation and vegetation cover. The affect of rock composition on soil texture and chemistry is greater because of the reduced rates of soil organic matter accumulation, decomposition, and transformation and removal of soluble salts. The sclerophyllous plant form, characterized by high lignin and cellulose content relative to protein content and by small, evergreen leaves, can be caused by shortage of water or by shortage of inorganic nutrients in the soil. This section details some of the similarities and differences in the geology and soils of the mediterranean scrub regions of California and Chile.

Tectonics, Geomorphology, and Lithology

Tectonics and lithology are of course independent of climatic influences. For comparisons of climatic effects in convergent evolution, then, it is fortuitous, but fortunate that the mediterranean areas of California and Chile have a similar geological history. Both are relatively young orogenic systems that gained their present gross morphology through violent up-thrusting in late Tertiary and early Quaternary times. These events took place on sites of ancient geosynclinal systems of the Mesozoic era that underwent profound disturbance and deformation leading to the creation of the southern Californian Batholith in the San Diego area and adjacent parts of Baja California, and the Andean Dioritic Batholith in Chile. The relative youth of tectonic instability of the mediterranean scrub regions of California and Chile is reflected in rather sharp, folded and faulted mountains and hills, often rising close to the coast with generally narrow and discontinuous coastal plains. Recent emergence and rapid denudation are reflected in strongly dissected scenery and terraced coastlands.

To a varying degree, climate influences geomorphic processes of landform development. Paskoff (1973) considered that the landforms of the mediterranean regions constitute a distinct group formed by both the present day processes of erosion and the changing climates of the Quaternary. Climatically, mediterranean regions are characterized by marked seasonality of precipitation accompanied by equally pronounced annual variations. Commonly the regime consists of a few widespread storms of high intensity that may lead to torrential flow. Consequently, the high runoff coefficient (the ratio between incident precipitation and that which runs off), resulting from concentrated precipitation, open vegetation, steep land surfaces, and thin soil, produces almost all of the erosion and deposition in these areas. This fact helps to explain why, in quantitative terms, the land surfaces in mediterranean regions approach similar inclinations, topographies, and soil depths. In addition valley profiles in such regions are commonly characterized by rectangular cross-sections and two-phase stream channels, as a result of the alternative down-cutting and aggradation of seasonal stream flows of high intensity. These generalizations are valid on a regional scale in both California and Chile, although local differences in relief, aspect, slope categories, and profile are to be found at the specific research sites.

Since the rock type affects soil characteristics in semiarid climates, the lithology of southern California and central Chile is important in evaluating soil differences. Reflecting their batholithic origin, which is associated with deep-seated igneous activity and granitization, the two primary sites are characterized generally by plutonic rock. However, the local petrology of the two sites is somewhat different, involving two distinct classes of igneous rocks differing in texture, chemical composition, and some mineral constituents. Quartz diorites occur at both sites, but the Chilean site exhibits, in addition, abundant light-green extrusive andesite with phenocrysts of plagioclase felspar highly weathered in kaolinite, a common clay mineral. In the Echo Valley area quartz diorite is the dominant parent material, at Fundo Santa Laura it occurs as an intrusive stock, cross-cutting the massive sequence of andesites. Within the Echo Valley area there are outcrops of gabbro, which is more similar in mineral composition to the Chilean andesite (Figure 3–14). Thus, the two climatically similar research sites provided the basis for an investigation of soils that developed from both similar and dissimilar parent materials under homologous climates. To further examine the influence of climate and parent material on soil development, rock and soil samples were also collected from a coastal and a montane site which are at approximately the same latitude as the primary sites in each country, but which have different climatic regimes than the primary sites. The comparisons between the

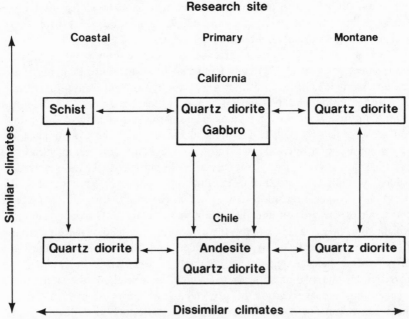

FIGURE 3–14. *Diagram identifying the predominant rock types at the coastal, primary, and montane research sites in southern California and central Chile.*

coastal, primary, and montane sites in each country are not completely parallel because the parent material is metamorphic schist at the California coastal site and quartz diorite at the Chilean coastal site and at the montane sites in both countries.

Andesite and quartz diorite, the two major rock types encountered from which the soil have developed, can be characterized as follows: both bear more plagioclase than potash feldspar, and tend to be slightly more basic than acidic; but whereas quartz diorite is intrusive, coarser grained, and more felsic, andesite is more mafic, with a fine-grained groundmass of pyroxene minerals and kaolinite. These differences in parent material are reflected in the soils of the two primary sites.

Soils

At the coastal, primary, and montane sites in both California and Chile, the soils generally are rather poorly developed, shallow in depth, commonly stony or gravelly throughout the profile, and neutral to slightly acid reaction. Soils developed in quartz diorite at Fundo Santa Laura are coarse textured, granular, loamy sands, while those at Echo Valley are sandy loams, with two to three times as much silt. Clay contents of these soils are similar, and both are permeable and well-drained. Since clay content is thought to increase with soil development (Zinke, 1973), the soils at the two primary sites are presumably of comparable maturity. Subsoils on quartz dioritic materials are generally less rudaceous, and commonly more yellowish-brown in color, reflecting in part the low ferrous mineral content of the parent material. Along the generalized transitions from rankers through brown earths to red loams, which characterizes soil development in other mediterranean areas, the soils occurring on the quartz dioritic materials fall somewhere between rankers and brown soils, depending upon the local topographic situation.

By contrast, soils developed on andesite at Fundo Santa Laura approach a red loam, and differ substantially from the soils on quartz diorite. The andesitic materials contain abundant kaolinite and weather to produce finer-textured soils than do the coarse-grained intrusive rocks. As a result the andesitic soils at the sites contain over three times as much clay as do the quartz diorite soils, and are thus classified as loams (nearly clay loams) and sandy clay loams. The subsoils associated with the andesitic parent rock are blocky in structure and reddish to reddish-yellow in color at bedrock. This coloration indicates the high ferrous mineral content of these rocks, and measurements indicated that free iron oxides were nearly four times as great in these soils as in the quartz dioritic soils. Outcrops of gabbro in the Echo Valley area also have relatively high metallic mineral content and are strongly rudaceous, but have intermediate concentrations of free iron oxides. Reflecting similar macroenvironments and parent material, the disjunct quartz-dioritic soils that occur at the matched Californian and Chilean primary sites are more alike than are the quartz-dioritic and andesitic soils at the primary site in Chile.

The influence of climate on soil development—both directly, through temperature and moisture, and indirectly, through vegetation cover—is evident by comparing soils derived from similar parent materials under different climates. In California and Chile the soils of the montane sites, which have developed from quartz diorite, are higher in organic matter and lower in pH and base saturation than the soils at the nearby primary sites. Thus, in certain characteristics at least, soils of similar parent rock are more alike at sites matched on the basis of climate but occurring on different continents than are soils of the same parent rock on adjacent, but climatically dissimilar, sites.

Other macroenvironmental factors, such as geographic position with respect to the ocean (past or present), may influence soil properties independently of climate or parent material. For example, the matched coastal sites, both of which are associated with marine terraces, have higher average sodium and magnesium concentrations than the adjacent primary or montane sites.

Soil Nutrients

Of particular importance to plant/soil relations are cation exchange properties, soil nutrient levels, and humus content (Table 3-5). Cation exchange capacities of all soils in both countries are low. In general, cation exchange capacity was found to increase with increasing clay content. Thus, in Chile soils developed from andesite have a cation exchange capacity more than twice that of soils on quartz diorite. Only soils from gabbro approached the cation exchange capacity of andesitic soils. In addition, the andesitic soils at Fundo Santa Laura have greater water holding capacity, due to small mean grain size, and higher plant nutrient storage capacity, due to the high cation exchange capacities. These characteristics undoubtedly are to some extent a reflection of the vegetative cover since the andesitic soils also are somewhat higher in organic matter than any of the quartz dioritic soils. All soils at the primary sites in California and Chile are characterized by a rather high calcium-magnesium complex and are at least three-fourths saturated with bases. Exchangeable bases (calcium, magnesium, sodium, and potassium) are most plentiful in the andesitic soils.

Deficiencies of nitrogen and phosphorus are common in mediterranean scrub regions, and such deficiencies have been tied to the evolution and migration of evergreen sclerophyllous (xeromorphic) vegetation (Beadle, 1966; Monk, 1966; Small, 1973). Research suggests that sclerophylly increases

TABLE 3-5 *Comparative Soil Characteristics at the Coastal, Primary, and Montane Sites in California and Chile.*

	Percent clay	Percent organic matter	pH	Cation exchange capacity (meq/100 g)	Exchangeable cation (meq/100 g)				Nitrate nitrogen (μg/g soil)	Available phosphorus (ppm)	Free iron oxide (%)
					Ca	Mg	Na	K			
Coastal											
Camp Pendleton											
0–0.3 m	18	2.2	6.0	21	9.0	6.0	0.9	0.3	14.2	2.1	1.1
0.3–bedrock	24	0.5	6.4	23	8.0	9.5	1.1	0.2	2.0	0.6	1.7
Papudo/Zapallar											
0–0.3 m	15	1.3	6.0	10	2.0	1.9	1.1	0.3	2.6	1.8	2.4
0.3–bedrock	23	0.7	5.2	13	2.0	4.9	5.4	0.2	2.3	2.1	3.0
Primary (ridgetop)											
Echo Valley											
0–0.3 m	10	1.4	6.8	13	7.4	1.5	0.6	0.2	4.7	1.6	1.2
0.3–bedrock	8	0.4	7.0	12	7.0	2.3	0.6	0.1	2.3	0.6	0.8
Fundo Santa Laura											
0–0.3 m	15	1.7	6.3	15	7.6	2.3	0.5	0.2	2.6	5.2	2.8
0.3–bedrock	32	0.8	6.5	30	17.3	9.5	0.8	0.3	1.0	0.4	2.1
Montane											
Mount Laguna											
0–0.3 m	10	1.8	6.5	12	6.8	0.9	0.6	0.4	2.7	14.4	1.3
0.3–bedrock	11	1.3	6.4	14	6.9	1.2	0.6	0.3	2.8	11.0	1.5
Cerro Roble											
0–0.3 m	5	2.5	6.3	11	4.1	0.3	0.6	0.4	1.7	17.5	1.0
0.3–bedrock	5	1.3	6.1	6	3.1	0.2	0.5	0.2	1.5	21.0	0.8

Source: Calculated from Bradbury (unpubl.)

with decreasing phosphorus levels in the soil. Interestingly, differences in these elements are found in the soils of the two primary sites. Soils at the Fundo Santa Laura site contain seven times as much phosphorus as the soils at Echo Valley. On this basis one would expect the dominant vegetative cover of the Chilean primary site to be more mesomorphic and perhaps of lighter mean plant weight. Presumably, the vegetation is partly responsible for these differences, since the two parent rocks do not reflect this degree of disparity, although kaolinitic soils have a greater fixation power for phosphorus (Allard, 1942).

In addition to its affect on plant form, phosphorus deficiency has also been considered limiting to nitrogen accumulation in soils, since it suppresses the development of nitrogen-fixing plants (Allard, 1942; Beadle, 1954). In Chile phosphorus contents of the soils on quartz diorite are higher than contents of soils on andesite, but nitrogen levels are similar on the two rocks. Since chaparral species in California, particularly *Ceanothus* spp., are nitrogen fixers, which may explain the overall higher nitrogen-nitrate content in the soils at Echo Valley than at Fundo Santa Laura. Zinke (1967) found that *Ceanothus* and scrub oak (*Quercus dumosa*) adds as much nitrogen per year as chamise (*Adenostoma fasciculatum*) depletes. Furthermore, the highest concentrations of nitrate at Echo Valley are found in the soils of the pole-facing slope and oakgrove; *Ceanothus* and scrub oak are common plants on the pole-facing slope.

Overall, the Chilean soils were found to have higher concentrations of soil nutrients. On the basis of soil alone, one might expect the dominant vegetation of the Chilean matorral to be, in general, taller, more mesomorphic, and with greater herbaceous cover than the vegetation of the California chaparral; however, the influence of other factors on vegetation, such as climate and man, must also be considered.

SUMMARY

The objective of this chapter was to outline the degree of similarity in macroenvironments of the mediterranean scrub regions of California and Chile, as two examples of regions with mediterranean climates. Since in these countries similarities in the pattern of the vegetation occur from the coast to the inland, the hypothesis was that the inland macroenvironments in the two countries would be more similar to each other than the macroenvironments of the inland and the coastal areas within each country. Aspects of micro-climate, geology, and soils were measured to indicate this degree of similarity. The essential aspects of climate, that constrain the vegetation, are related to the normal patterns of water availability, favorable temperatures, and light and to the historic variation in these availabilities.

As recently as 14,000 years ago both California and Chile experienced glacial advances in the Quaternary, which were associated with decreased temperatures and increased precipitation in the present mediterranean scrub regions. In the past 500 years climatic fluctuations of long duration have been less frequent in California than in Chile. Rainfall in central Chile has been above or below normal for periods of a 100 years or more, while in southern California such climatic shifts persist for only about 30 years. In both regions dry years are usually followed by dry years, but wet years are usually followed by a year of average precipitation. The amplitude of the variations in annual precipitation is greater in central Chile.

At present, the climates of both regions are controlled by the movements of the subtropical high pressure zones, which are evident as high pressure cells offshore. Precipitation is mainly caused by fronts, which penetrate the regions as the high pressure cells move equatorward in the respective winters. Precipitation increases inland because of orographic lifting of the air as it moves across the foothills and mountains. Precipitation from convectional storms is suppressed. Thus, precipitation, when it occurs, is widespread. In both countries saturated air or fog forms offshore due to the passage of warm, moisture-laden air across colder, nearshore water which has upwelled to the surface. The fog increases condensation and deposition of water. The latitudinal gradient in precipitation is less in California than Chile, and may contribute to the smaller amplitude of annual variability in California. Precipitation along the coast is about 250 mm in California and 350 mm in Chile. At 1 km elevation precipitation is about 550 mm in both countries and at 2 km elevation it is about 700 mm.

In California and Chile precipitation exceeds potential evapotranspiration only in winter months, when soil moisture can accumulate. Along the coasts precipitation exceeds potential evapotranspiration for only one to three months, but inland at 1 km elevation an excess occurs for three to five months. In the spring and summer, precipitation is less than potential evapotranspiration and soil drought occurs, for three to four months along the coasts and two to three months inland. Soil moisture is consistently present in the spring, but at amounts which vary according to the year to year variation in winter precipitation.

Temperatures in southern California are about $2°$ C higher than in central Chile. In both regions temperatures are less variable along the coasts than inland, but there is more diurnal and seasonal variability in the inland southern California area. Temperatures are warmer in winter and cooler in summer along the coasts than inland and are more favorable for plant activities along the coast in the winter. Inland in California, freezing temperatures occur in winter and hot, dry conditions predominate through the summer. Inland in Chile the temperatures are more moderate due to higher winds and increased penetration of marine air.

Global irradiance was greater in winter, less during summer in California. Along the coasts global irradiance was lower than inland due to coastal cloudiness Inland, global irradiance through the year was 250-560 cal cm^{-2} day^{-1} in California and 150-640 cal cm^{-2} day^{-1} in Chile.

Temperatures of similarly exposed soils were similar near the soil surface in both countries, but cooler at depth in Chile. When soil moisture is available temperatures are low, and when temperatures are favorable, moisture is unavailable for maximal biological activities in the soil.

The geology of the two regions is similar. Both are relatively young orogenic systems which gained their present gross morphology through upthrusting in late Tertiary and early Quaternary times. Recent emergence and rapid denudation show up in strongly dissected terraced coastlines. The similar patterns of rainfall and storm intensity, and of erosion, tend to produce slopes of similar inclination and soil depth, and valleys of similar cross section. Quartz diorites are common in both countries.

Soil profiles indicate similar periods for development and similar responsible agents. Clay contents are similar and soils at both inland sites are well drained. Soils of the montane sites showed higher organic matter, lower pH, and lower base saturation than soils at the inland sites in each country. Coastal soils in each country have higher sodium and magnesium concentrations than inland or montane sites. Californian soils showed lower cation exchange capacity, less phosphorus, and more nitrogen than the Chilean soils.

Thus, on comparing several regional and local environmental variables, the environments of the inland areas of the two countries are similar, and more similar than the environments of the inland and coastal areas in either country. In addition to such commonly reported similarities as the predominance of winter precipitation, moderate winter temperature, warm but not hot summer temperature, the regions also show similar solar and thermal irradiance, soil surface temperatures, soil moisture patterns, topography, and soil mineral composition.

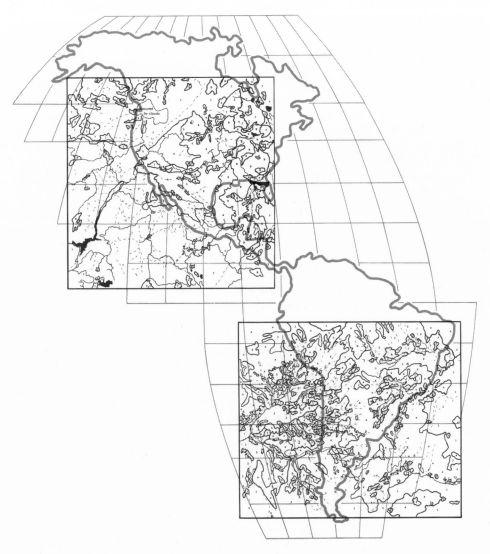

Man's Impact on the
Wild Landscape

H. Aschmann
C. Bahre

A broadening stream of archaeological evidence in California and the discovery and dating of the butchering site at Laguna de Tagua-Tagua in Chile (Montané, 1968) demonstrate that man has been part of the ecosystems of both regions for well over 10,000 years. After having been driven equatorward during glacial times, the climatic patterns and vegetation associations of the two regions achieved roughly their present distribution; and since that time the human component has been as much a part of the ecosystem as have the bees or the buzzards. It thus becomes pertinent to examine how human activities have affected other parts of the biota. During the past 10,000 years, and especially during the last 150 years, man's impact on his coresidents in regions of mediterranean type has increased at an accelerating rate. His numbers have increased, as have his cultural and technological capabilities. Though there is likely to be a positive linear correlation between man's numbers and the extent of his impact on his environing ecosystem, a far more complex picture is presented by the culturally conditioned variations in the species of plants and animals exploited, protected, or accidentally advantaged, and in how these actions were effectuated.

PRE-EUROPEAN DEVELOPMENTS

During the first eight millennia of the 10,000 years we are concerned with, human activities seem from the evidence available to have been comparable in the two areas. Thus, any variation in the effects of man's activities on the general aspect of the vegetation would have been due largely to innate or genetically controlled reactions to disturbance within the two floras. More recently, as the human populations of the two areas have increased in numbers and enlarged their technical capabilities, their treatment of the wild vegetation has become more disparate. The greatest variations have developed in the last century, far too short a time for major evolutionary changes in the plants themselves but quite long enough for man to have left an indelible mark on the appearance of the two vegetations and on the abundance of their component species.

Until well into the Christian era, fire was the most important agent by which man influenced the Chilean and Californian ecosystems. Whether deliberately spread or propagated accidentally from campfires or torches, it undoubtedly occurred more frequently than it would have without man's agency. The differential must have been greater in Chile, where summer thunderstorms are almost nonexistent. In any event, man-caused fires were a continuing feature of the environment in both regions during the entire post-Pleistocene period. The concentration of nitrogenous material around permanent or semipermanent campsites created distinctive microhabitats, attractive to weedy herbs. Probably the oldest and certainly the most abundant of these microhabitats in both Chile and California were shellmounds along the coasts.

The cultivation of crops in the parts of Chile with mediterranean climates long antedates European contact, whereas in California cultivation was delayed until the onset of European settlement in the late eighteenth century. The Chilean Indians even had three seed crops that were adapted to winter rains. In this respect, these crops were unique in the New World. They are identified in the chronicles as *teca, mango* (*Bromus mango*), and *madi* (an oil-yielding composite), and have been lost to cultivation because they could not compete economically with similarly adapted grains from the Mediterranean region itself (Keller, 1952: xlviii–xlix). Although some irrigated farming was carried on in valley bottoms from the Río Rapel northward, the basic farming system in the mediterranean regions of Chile involved clearing patches of flat or moderately sloping land by burning in the fall, planting tubers or seeds with the aid of a digging stick, and harvesting in the spring.

The best estimates of aboriginal population for the two areas of mediterranean climate at their respective contact times (the estimates for Chile for the mid-sixteenth century are extremely tentative) indicate that Chile had roughly one-and-one-half times as great a population density as California (Baumhoff, 1963; Medina, 1952: 157–160). The abundant and heavy-fruiting oaks of California seem to have nearly matched the nutritional contributions of the domesticated crops in Chile. Or it may be that the Indian population of central Chile simply had not built up to the maximal levels the land and technology would have supported. Since the peoples of both regions used fire regularly and had similarly semipermanent settlements, they probably had roughly equivalent impacts on the wild vegetation, even though one society was agricultural and the other was not. Slash-and-burn farming on interfluves in Chile would have effects comparable to the regular burning practiced by the California Indians. Domesticated llamas and alpacas were present in central Chile, but they were few in number, probably having been brought in with the Inca conquest no more than a few decades before the Spaniards arrived. There is no indication that these animals put grazing pressure on the vegetation as the European herd animals began to shortly after their arrival. From the Aconcagua valley northward in Chile, irrigated agriculture and permanent settlements were frequent. In California the historical and archaeological evidence indicates that the largest and most sedentary populations lived along the coasts.

PATTERNS OF WILDLAND USE IN CHILE

The conquest and permanent domination of the aborigines by Europeans was accomplished in the mediterranean parts of Chile by 1550; it did not begin in California until 1769. In both areas conquest was followed immediately by a precipitous decline in the human population. The native mortality, resulting primarily from disease and social disruption, destroyed three-fourths to four-fifths of the population within a generation and a half (Encina, 1954:131-133). The loss was not made up by immigration or mestization for nearly two centuries after the arrival of the Spaniards in Chile; for a century after their much later arrival in Southern California.

The European livestock introduced into Chile multiplied rapidly, putting pressure on the native grasses and herbs. Their replacement by the Old World weeds, such as wild oats and mustard in the level lowland areas, probably began in the sixteenth century and was essentially complete by the early nineteenth century, when the topic seems first to have been considered. Much less land had to be planted in colonial Chile than in precontact times because of the reduced population, and there was a tendency to concentrate cultivation in limited areas, often in places that could be irrigated. Grazing by the full set of European domesticated animals was the characteristic land use, in both hilly and relatively level districts. The cutting of woody plants for silver and copper smelting, as well as for structures and fuel, placed heavier pressure on trees and shrubs near towns and mines than had been known before the Conquest, though the most severe developments were deferred until the early nineteenth century. Particularly around major mining districts, man-made floral deserts persist to the present.

Rural land-tenure patterns, which developed in Chile during colonial times and have persisted until the present decade, constitute potent and distinctive determinants of the ways man has affected the wild vegetation. Except for designated cities and mining districts, virtually all of mediterranean Chile was assigned, along with its resident Indians, in *encomiendas.* Where the Indians were not actually farming, the encomendero could use the land for grazing. The encomienda was in fact an assignment of natives to the care of a conquistador, with the associated Indian lands as a secondary prize; as the institution was gradually discontinued, worthy Spaniards received land grants, often extensive ones. The Indian communities, thus freed of formal servitude, were awarded much smaller areas, often with one or two hectares of cultivable land per person, but sometimes with adjacent hill land for grazing small stock. North of Santiago, where nonirrigated farming is risky, a distinctive communal landholding institution, the *comunidad,* has come to control extensive hill lands that may be dry-farmed irregularly but are normally grazed as commons (Winnie, 1965).

During the nineteenth century the land grants that could be irrigated were developed as estates (*fundos*), often with more extensive adjacent uplands in the cordillera or the coast ranges, for grazing or other use. The resident labor force on the fundos, the *inquilinos,* received in exchange for their labor a hut, a garden, and the right to pasture small stock and gather wood on any wild lands the fundo held; similar rights were often sold or exchanged to *comuneros, minifundistas,* or landless day laborers resident in small rural towns (McBride, 1936:146-170; Góngora, 1960). All these elements of the rural labor force, collectively identified as *campesinos,* hold a common attitude toward wild lands: the vegetation is to be exploited for sale or use in any and all ways to eke out a difficult if not precarious existence (Bahre, 1974:141–165).

By the beginning of this century, if not considerably earlier, the woody vegetation on the communally held lands had been completely destroyed, except in the more inaccessible spots, by woodcutting, charcoal-making, clearing for dry-farming, and subsequent overgrazing, primarily by goats. The nonirrigable lands of the smallholders, notably in the Coast Range, were treated similarly.

Areas that might retain some wild vegetational character occur only on the larger fundos and haciendas, and in a few parks that constitute the public domain. For the landowner his wild lands yield significant rent, primarily through the grazing of cattle, and cutting and burning the brush to improve pasture is widely practiced. This thinning is so effective that few areas support enough fuel to carry hot and extensive wildfires, and a significant fraction of the shrubs survive the fires that do occur. The *espino* (*Acacia caven*) is a particularly enduring shrub or small tree occurring on relatively gentle slopes that are grazed but not farmed, and an espino savanna is perhaps the most extensive wild vegetational formation in mediterranean Chile. The extension of its range southward and over broad areas is largely the product of human activity.

The owner of an extensive property plays a much less direct part in other forms of vegetational exploitation. Permission to graze goats may or may not be granted to resident inquilinos as part of their traditional rights, or it may be sold for money or services to unaffiliated campesinos. Wood gathering for domestic fuel is commonly permitted residents, and cannot be effectively proscribed to neighbors. Campesinos living in the district are similarly permitted to collect wild plant material for food or as medicinals, and do so. Protection from such exploitation is afforded only by distance from a resident population.

Woodcutting and charcoal-making, or stripping the bark of the *quillay* (*Quillaja saponaria*) for its saponin, however, require explicit permission and some sort of payment. The payment is usually small, and a landowner might refuse permission without much economic loss. At the same time, where demand for fuel is or was great, as around irrigated settlements or mines, extirpation of the accessible woody vegetation has commonly occurred. The belief widely held among campesinos that cutting trees around a spring causes it to dry up (a belief that formed the basis for an 1872 law still in effect; Alfonso, 1909:12) has protected small patches of trees along minor watercourses on steep slopes, whereas the brush on the interfluves is severely abused and degraded (Bahre, 1974:177–178).

The survival of the most northerly remnant of the *Nothofagus* forest on Cerro Robles is a function of its distance from settlements. Where an unusually valuable vegetal resource persists in an isolated locality, e.g., the palms (*Jubaea chilensis*) behind Ocoa, campesinos from the closest settlements utilize them; in this case they camp among the palms and cut the trees and boil syrup from the sap. Protection of a rare plant against depredation by the poor has so far been politically impossible. Only distance and consequent low economic yield can keep the forager out of the wild lands, though an active protection policy may reduce his take to high-value, low-bulk items.

We can thus visualize the Chilean matorral (the vegetation association approximating the California chaparral) as a continuously used resource, with the intensity and completeness of exploitation a direct function of accessibility to the poor rural populations. Where the rural poor own the land, as in the comunidades, the wild vegetation has been extirpated, except on the steepest slopes. Where nonimpoverished persons or institutions control the wild lands, vegetational exploitation becomes more selective, requiring a more favorable ratio of value to bulk as inaccessibility increases. Idiosyncracies of large landholders have played a considerable role in determining the intensity of exploitation of the wild vegetation. Some have permitted goat grazing, woodcutting, or charcoal-making by their own inquilinos or outside campesinos as long as it would yield any rent at all. Most have allowed such activities, at least by their underemployed inquilinos, whenever there was a market.

The land-reform program of the past decade, which has expropriated virtually all large estates and turned them into cooperative farms known as *asentamientos,* has not yet had its full impact on the wild lands. Though the inquilinos who had been resident on each estate now operate a farm, the government advisor assigned to the asentamiento might encourage more conservative use of the vegetation; at the same time he is under heavy pressure from the members to permit any activity that promises them income.

Fundo Santa Laura, the IBP's primary site in Chile, is located in rough terrain just east of the Cuesta de Limache in the Coast Ranges northwest of Santiago. There is some rain-shadow effect here, and dry-farming—extensive on the west side of the Cuesta—is restricted to favorable soils and slopes. Though there are springs on the fundo, their flow is insufficient for significant irrigation, and only limited areas are level enough for dry-farming. These areas were cleared and cultivated, but mining, smelting, grazing, charcoal-making, and woodcutting were the principal sources of income for the fundo until 1959. Numerous charcoal ovens are scattered over the area. In the period immediately prior to that date, 9 to 11,000 kg of firewood and 5,000 kg of charcoal (requiring perhaps 20,000 kg of wood for its preparation) were sold off the fundo monthly. Still earlier, a copper smelter on the fundo must have consumed even greater quantities of wood.

In 1959, the fundo's owners instituted a vegetational-protection policy, halting the sale of firewood and charcoal. Dry-farmed fields and orchards were maintained until 1965, and the resident labor force used the wild vegetation for its own needs. Since that date only a single caretaker and his family have remained on the fundo, maintaining a small orchard and a garden on the lower part of the property near the highway. Fundo Santa Laura is close to the comunidad of Tiltil, with its large, poor, and underemployed population. It would be impossible to keep the comuneros from gathering medicinal plants and dead wood on the nearly unoccupied fundo, and several recent fires, the scars of which are still visible, probably resulted from such clandestine operations. These fires, however, were restricted to a few scores of hectares.

Immediately west, i.e., upslope, from the fundo headquarters three intermittently flowing quebradas join. The lower parts of their interfluves are smooth, and slope moderately toward the east. Until about 1965 these surfaces were farmed in potatoes, wheat, and alfalfa, and there were plantings of vineyards and almonds on about 40 hectares. Spring flow was diverted for a little irregular irrigation. The IBP weather station is located on the central interfluve, an area of grass, herbs, and scattered bushes that represents major vegetational disturbance. In contrast, the sharply incised quebradas are heavily wooded with matorral plants that often reach tree proportions, up to ten meters in height. The steeper south-facing slopes carry fairly heavy stands of matorral; the north-facing, thinner stands. Most of the larger shrubs have resprouted from cuttings, and stumps where branches have been hacked off surround their bases (Bahre, 1973:4).

The forested character of the quebradas and the localities around springs results from the protection afforded by the well-established myth that cutting trees around a spring or stream will cause the water source to dry up.

PATTERNS OF WILDLAND USE IN CALIFORNIA

The European occupation of the mediterranean parts of California began in 1769. As in Chile, the occupation resulted in a rapid decline in the native population (Cook, 1943). Further, to the extent they were able, the missionaries assembled the Indians from their numerous independent bands or *rancherías* into a few permanent settlements where irrigated agriculture and fixed-base grazing could be pursued. Thus, extensive upland areas that had been moderately populated and exploited in aboriginal times, at least seasonally, were essentially abandoned for a few decades. And although man-set fires in the chaparral and oak-grass parklands were not eliminated, they probably declined substantially in frequency.

Extensive land grants to individuals had begun prior to 1822, the year marking the beginning of the Mexican period, but they accelerated in the decades that followed. The lands thus granted, situated some distance from the missions, were used principally for grazing cattle and horses. By the middle of the century Old World grasses and annual herbs had largely replaced the native species. Very likely, this amazingly rapid vegetative displacement was the result, as it was in Chile, of the superior capacity of the Old World annuals to resist heavy grazing and trampling by large domestic stock. The extent to which fires were set in range land during the Spanish, Mexican, and early American periods, as compared with aboriginal times, is unclear. Some early reports condemn the Indians for deliberately setting fires, but burning to remove brush and improve the next year's grass crop was not unknown to European stockmen and in fact was widely reported in the mid-nineteenth century in Northern California (Shantz, 1947:84-109). Some fires, certainly, were accidentally set; and the ranchers were too few to have effectively suppressed fires of whatever origin.

Following the suppression of the missions in 1833, some Indians reverted to their native ways, though in reduced numbers, living by sufferance on lands claimed by the ranchers who continued to dominate the wild landscape (Shipek, 1968). During the droughts of the 1860s, most of the Mexican land grants came into Anglo-American possession, and in the 1880s a sudden influx of immigration into Southern California brought about a rapid expansion of both irrigated agriculture and dry farming, the latter increasing to nearly its present extent.

A growing demand for water for irrigated agriculture and urban use developed a concern for watershed protection. Especially troublesome was the clogging or destruction of water-control works by the debris from the accelerated erosion that followed brush and forest fires. In response to this concern, the National Forest system was expanded to cover some 10 percent of the total land area of southern California, beginning with the establishment of the San Gabriel Forest Reserve in 1892 (U.S. Forest Service,1962). These forests occupy the highest lands of the region, areas that had not been granted as ranches by Mexico and had not been claimed for farming by homesteaders. They include almost all the native coniferous forests in the region and even more extensive areas of chaparral. Most areas of oak-grass savanna had already been taken over by ranchers and were not included in the National Forests. ·

Down almost to the present, enthusiastic fire prevention or suppression has been the watchword of the National Forest Service, and its influence has extended to the privately owned wildlands surrounding the National Forests. Although increasing population and greater recreational use of wildlands have led to an increased incidence of accidentally set fires, fire suppression has become increasingly effective. And although fires still occur, the effectiveness of modern firefighting is remarkable. It can best be characterized by how frequently the typical area of brushland or rangeland burns; this figure may run to about once in forty years, whereas in aboriginal times it was probably about once in eight years (Dodge, 1972). In northern California, burning by ranchers to keep down brush and encourage grass has gained some degree of acceptance against the opposition of the National Forest System (Shantz, 1947:84-109); but in southern California, the deliberate burning of wildlands continues to be almost completely proscribed, even under controlled conditions.

The fires that do occur in mature or overmature chaparral actually constitute a new element in the environment. Dead branches, erect and well aerated, dry in the long summers and form a superb fuel; and over forty years they attain a vastly greater volume than over eight years. Wildfires of extreme intensity and rapid spread, such as occur today, can cover thousands of hectares in a few days, killing thick-barked oak trees and destroying the roots of shrubs that would sprout after less intense fires. A fire in Chile that consumes 500 hectares of brush before it dies or is controlled is unusual. Fires more than an order of magnitude larger and much hotter—especially those driven by the Santa Ana winds—occur in California (Dodge, 1975).

How the occurrence of most of the individual species of shrubs is affected by fires of differing intensities is not understood, though some fire adaptations are evident. Root- or stump-sprouting following moderately intense burns is characteristic of several species in California and Chile, but the most intense burns in California often preclude it. The coniferous forest at the upper border of the chaparral seems to be especially vulnerable to a wildfire moving into it. On all the mountain ranges of southern California, the lower timberline is known to have retreated upward in historic times, in many instances following specific fires. The pinyon pines are particularly affected, and in southern California they have become almost restricted to the lower and desert side of their natural range, areas too rocky and sparsely vegetated to carry hot fires (Dodge, 1975). A number of species seem to seed effectively only on mineral soils exposed by burning. The nearly solid stands of chamise (*Adenostoma fasciculatum*) in the California chaparral seem to have no counterpart in Chile, and such stands occur in places that have histories of frequent fires.

The specific fire history of the Echo Valley area, chosen by the IBP (Origin and Structure of Ecosystems Project) as its primary site in California, is illustrative. The area is just within the Cleveland National Forest, and records maintained by the Fire Laboratory of the U.S. Forest Service in Riverside, California, show that the area was last burned in August 1950. That fire was of considerable intensity and burned over 8,000 hectares. The oak woodland in the valley bottom just below the site, along with some other valley-bottom groves within the area of the burn, survived. The U.S.G.S. Tule Springs quadrangle (scale 1:24,000), prepared in 1954 and field-checked in 1960, shows the burned area as unwooded, the valley-bottom oak groves as forest, and the unburned chaparral-covered slopes as brush-covered. By 1971, when the site was chosen, the chaparral had recovered and was in vigorous growth. It is actually a young stand in the terms of southern California chaparral stands in this era, but by now it is thick enough that cattle being grazed in the forest stay in the bottomlands, finding little forage on the hill slopes.

At elevations 100 to 200 m higher, some 5 km to the north of the IBP site, in an area that was burned at the same time, vegetation has recovered more rapidly, and the chaparral stands are taller and denser. Some wooded slopes, dominated by live oaks not killed by the 1960 fire, remain forested. A reasonable explanation, though the available records are inadequate to document it, is that the higher areas had received enough summer rain to reduce the intensity of the fire. To the south of Highway 80, 8 km away, the brushland was burned in the great fire of late September 1970. Here, very little recovery has occurred; grass and rocky, exposed soil dominate the landscape.

A kilometer to the east of the IBP site several ranches are established, and the valley bottom is maintained in grass for grazing purposes. There are no other areas in the immediate vicinity where major cultural disturbance is evident. In contrast to the analogous areas in Chile, the California brushlands in this area are either fully disturbed and under management or show only the effects of particular great fires, modulated by local variations in exposure or edaphic conditions.

The extensive chaparral-covered areas in California experience very little grazing pressure, though the grass and oak parklands are grazed by cattle as intensely as the grasses can bear. Except at its margins the chaparral affords little feed for cattle, though the coastal sage scrub at lower elevations has some value as pasture. Goats that could subsist on the chaparral are not part of the California economy, and sheep would lose their wool going through the brush. In some compensating measure the larger mammalian herbivores, notably deer and rabits, are far more abundant in California than in Chile. Indeed, in Chile the only such herbivores abundant enough to affect the vegetation are small rodents such as the *degu* (*Octodon degus*), which supposedly eats enough of the seeds of the Chilean palm (*Jubaea chilensis*) to impede its reproduction. Although sport hunting of deer and rabbits is popular in California, it is controlled to maintain a maximum population, probably a greater one than that in aboriginal times. Whether the virtual faunal desert of Chile is the product of centuries of indiscriminate hunting or of other causes is not known (Hernández, 1970; Fuenzalida Villegas, 1965; González Rodríquez, 1966).

The management of the wildlands of southern California involves a curious mixture of rationality and affluent profligacy. The publicly owned lands are valued chiefly for recreational purposes, both of a transient sort and as second residences on leased land. Limited grazing and lumbering are also permitted. Fire is regarded as anathema to recreational use and is a genuine threat to the resort communities. Severe erosion and the concomitant and more serious depositing of debris in adjacent lowlands following burns are also recognized problems. Property size in the privately owned areas varies broadly, ranging from plots of one to two hectares to ranches of thousands of hectares. Grazing operations are undertaken only by a relatively small number of large landholders who also lease grazing rights in the National Forests and on the adjacent smaller properties, many of which are abandoned dry farms. Land values and taxes on wildlands are ridiculously high in relation to their grazing return, and few of the owners gain significant returns in their total operation. Consequently, although the large properties are grazed in rational fashion, there is relatively little effort to intensify production to a maximum by increasing labor input.

Land is held in part for the owner's recreational hunting and riding, in part as a tax-deductible speculation. Ranches are also being broken up into 1- to 5-hectare properties or even small lots. In places low enough that steep slopes are frost-free, sprinkler-irrigated citrus and avocado orchards are being developed, but the main thrust is toward retirement, weekend, or exurban home sites. The heavy chaparral or woodland is an amenity, and the prospect of wildfire is a physical as well as an esthetic threat. A tightening circle is developing in which the wildland residents demand stronger fire protection, and in receiving it create an overmature brushland with growing quantities of dead fuel. Relatively infrequently, but with devastating effects, an uncontrollable fire breaks out under Santa Ana wind conditions with high temperature, low humidity, and high winds. Steps to modify this pattern will be difficult to take.

CONCLUSIONS

Until about a century ago, then, the wild vegetation of the mediterranean parts of Chile and California experienced in the same sequence roughly the same pressures from human activities. In both regions essentially all the land suitable for agriculture has since been taken under permanent management, and some marginal lands on slopes that were once farmed have reverted to wildlands. The treatment of wildlands, however, has progressively diverged. In Chile the pattern is one of maximum exploitation of vegetal resources, limited for the most part only by accessibility. Where the resident poor control the land use the combination of woodcutting and overgrazing has reduced the wild vegetation almost to a desert. And where the lands are in large holdings, selective exploitation has raked the flora for anything valuable enough to carry to market.

In southern California there has been a retreat from the direct exploitation of the wild vegetation. With rising living standards and labor costs, the selective exploitation of the vegetation by woodcutting, herb gathering, or intensive grazing of small properties has become less and less feasible. Increasingly, people who earn their livings in urban centers have come to prize the beauty and recreational virtues of undisturbed wildlands. Fire protection and suppression have therefore become the dominant factor in land management, and occasional severe wildfires have been the unintended result.

The aspect of the mediterranean scrub of California and Chile as now observed differs substantially in two gross ways. Except along watercourses and at springs, the Chilean form is more open; shrubs are of comparable heights but stand farther apart. Within plots of limited extent the floristic diversity in Chile is nearly double that in California (Parsons and Moldenke, 1975).

The continuous selective attack by man, coupled with the absence of fully destructive fires that would be followed by a vegetational succession of uniform age, can account for the openness of Chile; and the open stands, with plants of varying ages, offer more microhabitats appropriate to a more diverse range of species. In large measure, man's different practices in the two regions explain the different vegetational character of the two regions.

The Producers-Their Resources and Adaptive Responses

H. A. Mooney   A. Hoffmann

J. Kummerow   R. I. Hays

A. W. Johnson   J. Giliberto

D. J. Parsons   C. Chu

S. Keeley

The previous chapters have documented the degree of environmental similarity between southern California and central Chile. It was concluded that not only are the broad aspects of the extant climates similar, but the climatic changes that have occurred from the Pleistocene to the present in the two regions have been generally comparable in direction, magnitude, and timing. Furthermore, equivalent orogenic and soil-forming processes have led to similar land forms and soil types in the two regions.

It was further concluded that the taxonomic affinity of the woody floras of California and Chile is essentially nil. Thus the two mediterranean regions we studied offered a unique opportunity to determine the influence of environment on the evolutionary molding of plant forms and community structure, since these regions represent rather close replications of a distinctive environmental complex and have experienced essentially no genetic exchange. Moreover, the land-use patterns of the two study regions were quite comparable in the early days of European settlement, though they have diverged to such an extent since that today there is not possibility of matching study sites precisely in this respect. The differences have, however, been documented (see Chapter 4), and they, also, provide a framework of interpretation for our biological observations.

THE MATTER OF CONVERGENCE

Thus, we knew at the outset of this program that in general terms the vegetations of California and Chile were very similar in gross appearance, or physiognomy. This in itself is already an indication of the evolution of convergent forms in comparable environments. What we ask now is whether the convergence in form so evident at the gross-vegetation level is indeed a reflection of similarities in various parts of these systems. In other words, to what level of organization and detail is convergence manifest?

The degree to which there is a correspondence between not only the forms of the plants but also their environmental responses sheds light on the extent to which the adaptive responses to a given environmental situation may be limited in number. That is, when these environments are repeated in various parts of the world, will similar forms evolve independently, in spite of regional differences in evolutionary history?

Our study was directed primarily toward detailing the structural similarity between the vegetations of mediterranean California and Chile, and relating these structural attributes to environmental gradients. This approach essentially documents the outcome of an evolutionary experiment. Environments were matched as carefully as possible at two points (the primary sites) on two separate continents that support distinct floras, and the community structures were quantified. And by studying additional sites in close proximity to the primary sites, we could examine taxonomically related plants growing under climatically dissimilar conditions. We could thus test the hypothesis that greater similarities should occur at the matched sites on the two continents than at any two adjacent but climatically dissimilar sites on the same continent.

If convergence had indeed occurred—which seemed, on the strength of our preliminary observations, to be likely at least to a certain degree—we wanted to know what particular mechanisms give the extant plant types the selective edge—and presumably establish theirs as the most adaptive of possible strategies. Why is a given form-functional type more adapted for the constraints of a mediterranean climate than another? Why do certain forms recur again and again in distantly related stock within this climatic regime?

In order to answer these questions, we studied certain functional attributes of the plants related to phenological development, the schedule of carbon gain, and the apportionment of carbon to various alternative requirements or needs. Because of the time involved in certain of these measurements, they were generally centered on plants of the primary sites. In these cases, the test of convergence was a comparison of our results on plants of the matched primary sites in California and Chile with the literature on similar processes in plants from dissimilar climatic types.

Finally, some processes were studied in such detail that they could be examined only in a few or even in single species, which were then taken as representative of the evergreen adaptive mode. Again, we made comparisons between our results and those reported in the literature for similar taxonomic groups in different climatic regimes and for dissimilar taxonomic groups in the mediterranean climatic region.

From these various observations and comparisons we hope to answer the questions whether convergence had indeed occurred and, if so, to what extent and by what mechanisms.

VEGETATION STRUCTURE

Geographic Trends and Contrasts

A broad view of the relationships between plant form, vegetation structure, and environmental type is given by comparisons of the vegetations of two sets of climatically matched sites, four in California and four in Chile. These sites represent a gradient of decreasing aridity, from desert to coastal mediterranean, interior mediterranean, and, finally, montane climatic types.

Floristic comparisons of these sites show that there are virtually no generic affinities between any two matched sites, but between adjacent climatic types within each of the continents (Figure 5-1) there are strong relationships (Parsons and Moldenke, 1975). In spite of this evident lack of common evolutionary history, the features of the vegetation are more alike between the matched sites on different continents than between the adjacent, climatically dissimilar sites within either area. For example, the quantitative distribution

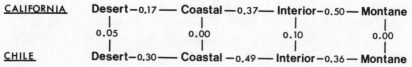

CALIFORNIA   Desert—0.17 —— Coastal—0.37— Interior—0.50— Montane

FIGURE 5-1. *Generic affinities of the woody vegetations of various climatic vegetation types in southern California and central Chile. Each similarity coefficient equals the number of genera common to the two sites divided by the sum of genera found at both sites (Parsons and Moldenke, 1975).*

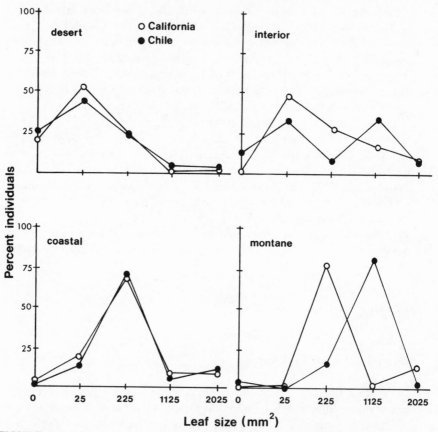

FIGURE 5-2. *Percentage of individuals of shrub species falling into various leaf-size categories, at matched sites in California and Chile (Parsons and Moldenke, 1975).*

of plants having various leaf sizes (Figure 5-2) and leaf retention (Figure 5-3) at the climatically matched stations is virtually identical. At both the xeric and the cold-mesic ends of the climatic gradients, the majority of the dominant plants are deciduous, whereas, at the interior mediterranean stations, evergreen shrubs predominate. Succulent species are important only at the

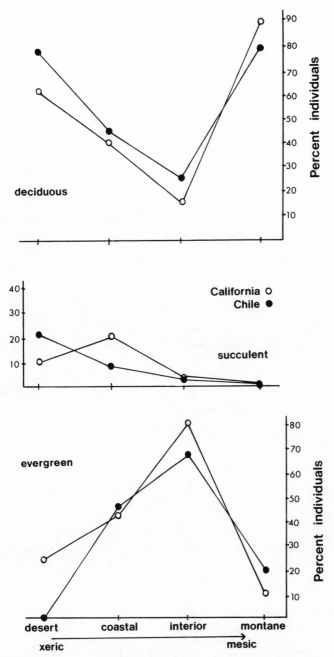

FIGURE 5-3. *Percentage of shrub individuals with deciduous, evergreen, and succulent habits at four pairs of matched sites in California and Chile (Parsons and Moldenke, 1975).*

xeric sites. The desert shrubs have predominantly small leaves, whereas larger-leaved species are found in greater abundance in the most mesic habitats. Spiny individuals are prevalent at both the desert and the interior sites (Table 5-1).

The plant-growth forms of the matched stations are also quite comparable (Table 5-1). For example, the most arid sites have the highest percentage of annuals (therophytes), the coastal sites the lowest percentage of shrub or tree species. All stations, however, exhibit a life-form spectrum characteristic of arid regions.

Many vegetation features of the matched stations are quite dissimilar, although the general trends are comparable along the two aridity gradients (Table 5-1). The percentage of woody plant cover decreases with increasing aridity. However, in all cases, the actual cover is much greater at the Californian stations than at the analogous Chilean sites.

The number of plant species, total and woody, is least at the desert sites in both continents. In all cases, though, the number of species is much greater in Chile than in California. The number of woody individuals is least at the desert sites in both continents.

In both California and Chile the woody plants show a shift from shorter to taller individuals with increasing habitat moisture (Figure 5-4). Generally, there is a greater diversity of height classes in Chile than in California at any given station.

The degree of cover by a single plant species or by just a few species is much greater at the Californian sites than at their Chilean counterparts (Figure 5-5). For example, at the California primary site (Echo Valley) two species, *Adenostoma fasciculatum* and *Ceanothus greggii*, constitute over 75 percent of the relative cover, whereas at the Chilean primary site (Fundo Santa Laura) no less than five species account for this much of the relative cover (Table 5-2). A similar relationship favoring higher California dominance—i.e., concentration of the cover in a few species—holds at the arid sites.

The broad comparisons between plant-community features along a large environmental gradient indicate a high degree of convergence in plant forms at comparable climatic types in the two continents, with a somewhat lesser degree of convergence in such vegetation-structural features as cover and diversity.

Specific Site Comparisons: Altitudinal and Topographic Vegetation Patterns

An even more convincing demonstration of the convergent aspects of the vegetations of California and Chile is seen in comparisons of equivalent elevational (350-1,850 m) and topographic positions within a single climatic type in the two areas. Even within the fine scale of microclimatic variation found

TABLE 5-1 *Comparisons of the Vegetation Characteristics of Climatically Matched Sites in California and Chile (aridity decreases with progression from desert to coastal to interior sites)*

	Site					
	Coastal desert		Coastal mediterranean		Interior mediterranean	
Vegetation characteristic	Chile	California	Chile	California	Chile	California
Total species	76	39	109	65	108	44
Total woody species	25	16	30	21	45	29
Total woody individuals	364	878	1,667	3,400	2,070	1,960
Total percent woody plant cover*	32.1	58.2	48.4	99.5	59.3	195.3
Percentage shrub individuals with spines	41	68	9	9	32	25
Growth-form composition: percent of *total* flora						
Phanerophytes	23	18	12	17	24	41
Chamaephytes	11	23	15	19	14	16
Hemicryptophytes	11	15	17	20	20	18
Geophytes	7	0	9	3	6	11
Therophytes	49	43	46	41	36	14

Sources: All data are from Parsons and Moldenke (1975) and represent total values from ten, 100 m² replicates at each station, except *, which are from Keeley and Johnson (unpub. data) and represent averages for all slopes at all stations. The desert sites in the two analyses differ. Plant overlap accounts for the cover value in excess of 100 percent.

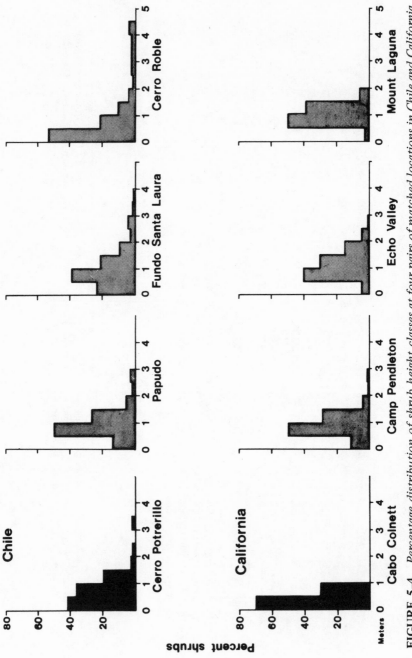

FIGURE 5-4. *Percentage distribution of shrub height classes at four pairs of matched locations in Chile and California, average of all slopes (Keeley and Johnson, unpublished data).*

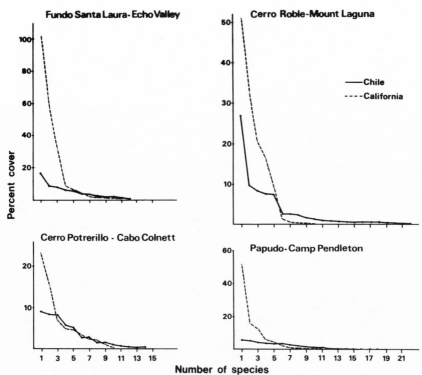

FIGURE 5-5. *Numbers of shrub species attaining various cover values at four pairs of matched stations in Chile and California (Keeley and Johnson, unpublished data).*

on a single mountainside, there is a comparable gradient of convergent forms on the two continents. For example, woody plants with similar leaf types generally predominate at the same point on the topographic-elevational gradient in both California and Chile (Figure 5-6). Plants with large evergreen leaves are present at the low-elevation, mesic sites, and those with winter-deciduous mesophytic leaves occur in moist, high-elevation habitats. Drought-deciduous mesophytic and/or small, sclerophyllous evergreens, which represent alternative drought-adaptive strategies, are located at the low-elevation, xeric habitats.

The same relationships evident with individual characters hold when we compare character complexes. Parsons (1976) examined all of the principal woody species in an elevational gradient extending from the primary sites, dominated by evergreen shrubs, upward into and downward into xeric subshrub communities, in both California and Chile. Utilizing seventeen separate plant-structural characters, he recognized nine character groupings (Figure 5-7). Plants of each of these groups have comparable distributions in California and Chile. For example, the narrow-leaved open shrub types

TABLE 5-2 *Average Absolute and Relative Percentage Shrub and Tree Cover from Three Pairs of Climatically Matched Sites in California and Chile*

	Shrub cover	
Shrub species	Absolute percent	Relative percent
California: mediterranean interior (primary site, Echo Valley)		
Adenostoma fasciculatum	101.86	52.15
Ceanothus greggii	49.69	25.44
Quercus dumosa	14.39	7.37
Arctostaphylos glauca	9.67	4.95
Eriogonum fasciculatum	5.33	2.73
Arctostaphylos glandulosa	4.46	2.28
Rhus ovata	2.13	1.09
Salvia clevelandii	2.05	1.05
Cercocarpus betuloides	1.77	0.91
Ceanothus leucodermis	0.81	0.41
Yucca whipplei	0.77	0.39
Quercus engelmannii	0.71	0.36
Rhamnus crocea	0.48	0.25
Trichostema parishii	0.35	0.18
Rhamnus ilicifolia	0.23	0.13
Salvia apiana	0.23	0.12
Penstemon ternatus	0.22	0.11
Artemisia californica	0.13	0.07
Lonicera subspicata	0.01	0.01
Total percent cover	195.31	
California: mediterranean coastal (Camp Pendleton)		
Salvia mellifera	53.43	53.86
Rhus laurina	16.58	16.71
Artemisia californica	13.31	13.42
Eriogonum fasciculatum	7.15	7.21
Rhus integrifolia	5.13	5.17
Cneoridium dumosum	2.51	2.53
Quercus dumosa	0.94	0.95
Yucca whipplei	0.31	0.31
Galium nuttallii	0.08	0.08
Dudleya farinosa	0.03	0.03
Opuntia occidentalis	0.03	0.03
Total percent cover	99.50	
California: coastal desert (San Telmo)		
Franseria chenopodiifolia	17.63	30.28
Rosa minutifolia	15.73	27.01
Agave shawii	7.90	13.57

TABLE 5-2 (Continued)

| Shrub species | Shrub cover | |
	Absolute percent	Relative percent
Simmondsia chinensis	7.25	12.45
Machaerocereus gummosus	4.00	6.87
Harfordia macroptera	1.91	3.28
Bergerocactus emoryi	1.16	1.99
Dudleya ingens	0.83	1.43
Eriogonum fasciculatum	0.83	1.43
Euphorbia misera	0.41	0.70
Mammillaria dioica	0.33	0.57
Echinocereus maritimus	0.25	0.43
Total percent cover	58.23	
Chile: mediterranean interior (primary site, Fundo Santa Laura)		
Cryptocarya alba	16.79	28.32
Colletia spinosa	8.15	13.75
Satureja gilliesii	7.87	13.27
Lithraea caustica	6.58	11.10
Trevoa trinervis	5.02	8.47
Colliguaya odorifera	4.21	7.10
Gochnatia fascicularis	3.13	5.28
Kageneckia oblonga	2.17	3.66
Baccharis linearis	1.75	2.95
Pondanthus mitique	1.21	2.04
Teucrium bicolor	0.54	0.91
Retanilla ephedra	0.39	0.66
Quillaja saponaria	0.32	0.54
Haplopappus longipes	0.27	0.46
Muhlenbeckia hastulata	0.17	0.29
Puya berteroniana	0.17	0.29
Mutisia berteriana	0.16	0.27
Trichocereus chilensis	0.10	0.17
Ephedra andina	0.08	0.13
Ribes trilobum	0.07	0.12
Baccharis macrei	0.06	0.10
Talguenea quinquinerva	0.06	0.10
Proustia pungens	0.02	0.03
Total percent cover	59.29	
Chile: mediterranean coastal (Papudo)		
Baccharis concava	8.20	16.94
Lithraea caustica	6.97	14.39
Baccharis linearis	6.29	12.99
Bahia ambrosioides	5.61	11.59
Puya chilensis	5.50	11.36
Lobelia tupa	4.38	9.05

TABLE 5-2 (Continued)

Shrub species	Shrub cover	
	Absolute percent	Relative percent
Fuchsia lycioides	3.45	7.13
Lepechinia salviae	2.36	4.87
Podanthus mitique	1.74	3.59
Escallonia pulverulenta	1.18	2.44
Cryptocarya alba	1.14	2.35
Chusquea cumingia	0.55	1.14
Eryngium paniculatum	0.43	0.89
Adesmia arborea	0.21	0.43
Eupatorium glechonophyllum	0.14	0.29
Haplopappus polyphyllus	0.12	0.25
Ribes trilobum	0.12	0.25
Muhlenbeckia hastulata	0.03	0.06
Total percent cover	48.42	
Chile: coastal desert (Cerro Potrerillo		
Puya chilensis	5.99	18.65
Flourensia thurifera	5.67	17.66
Oxalis gigantea	5.11	15.91
Heliotropium rosmarinifolium	3.94	12.27
Llagunoa glandulosa	3.79	11.80
Baccharis linearis	1.60	4.98
Fuchsia lycioides	1.27	3.96
Hoffmanseggia falcaria	1.19	3.83
Eulychnia spinibarbis	1.18	3.67
Cassia coquimbensis	0.88	2.74
Adesmia arborea	0.55	1.71
Proustia pungens	0.55	1.71
Bahia ambrosioides	0.18	0.56
Ephedra andina	0.12	0.37
Centaurea chilensis	0.08	0.25
Colliguaya odorifera	0.01	0.03
Total percent cover	32.11	

Source: Keeley and Johnson (unpublished data).

are found in the most exposed habitats and are replaced by broad sclerophyll shrubs in more mesic sites.

Furthermore, Parsons found by this analysis that in many cases the closest morphological analog to a given species was not the taxonomically most closely related species on the same continent but rather a species in a different family on a different continent occupying a similar habitat—strong additional evidence that phylogenetic history can be preempted by environmental selection.

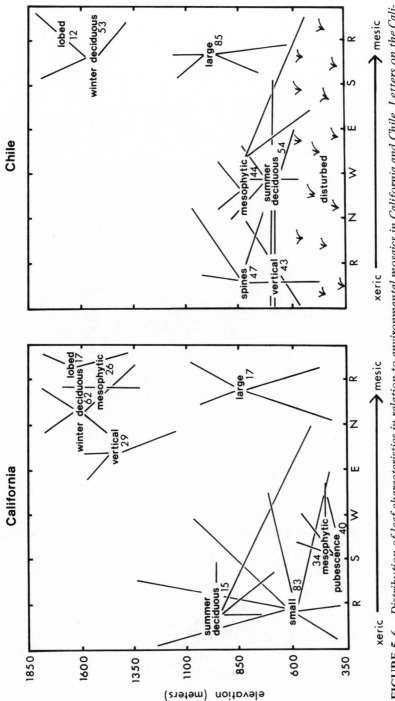

FIGURE 5-6. *Distribution of leaf characteristics in relation to environmental mosaics in California and Chile. Letters on the California graph indicate ridge, south, west, east, north facing slopes and ravine. The comparable topographic positions are reversed in the Southern Hemisphere plot. The point of maximum occurrence of the various features is determined on the basis of percentage plant cover; areas within about 80 percent of that maximum lie within the lines radiating outward (Parsons, 1975). The lowest elevations in Chile were highly disturbed and thus not included in the comparison.*

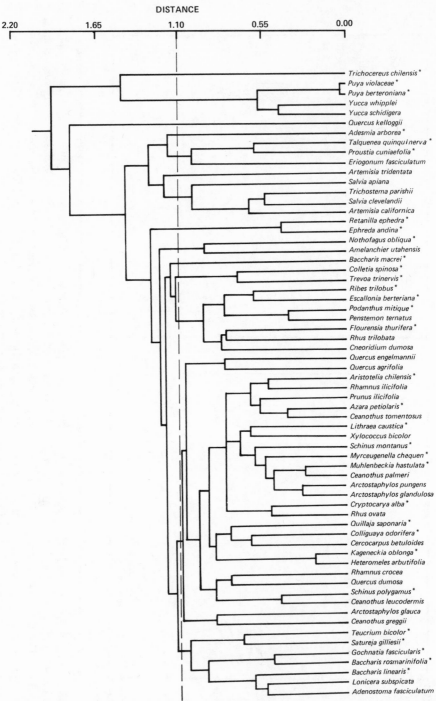

FIGURE 5-7. *Phenogram for dominant shrub species at the Chilean (*) and Californian primary sites (Parsons, 1976). Coefficients of distance are summarized in the form of a phenogram for 64 dominant shrub species from the matorral and*

Primary-Site Comparisons

The most detailed studies of the vegetation were focused on the primary sites, centering at about 1,000 m elevation. The results of these studies provide the finest resolution of comparison between the Chilean and Californian evergreen-sclerophyll plant communities.

The topography of the two regions is quite comparable, with a diversity of ravines and slope faces. The ravines are forested in both regions (Figure 5-8), and sampling was concentrated on the slopes.

The growth-form composition of the woody vegetation is quite similar in the two regions, with virtually identical representation of the major growth forms (Table 5-3). However, in Chile there are *additional* growth forms, such as spinose- and chlorophyllous-stemmed shrubs, that are absent in the California chaparral.

Table 5-4 details the cover distribution of the woody species at these sites. As indicated earlier, plant cover and species dominance are greater in California than in Chile, floristic diversity less. In Chile the tree *Cryptocarya alba* is an important component of the slope vegetation; there is no tree counterpart on the slopes in California. The greater height diversity of the woody plants at the Chilean primary site, vis-à-vis the Californian (Figure 5-4), is illustrated diagrammatically in Figure 5-9.

One of the most dramatic differences between the vegetations of the two primary sites lies in the degree of herb cover. In Chile, within the unburned slope vegetation, Keeley and Johnson (1976) encountered over 60 percent herb cover, constituted by over forty species (Table 5-5). More than one half of these herb species are perennials, but a substantial portion of the actual cover is composed of introduced Mediterranean annuals. These introduced annuals are confined to areas away from the shrub canopies, whereas native herbs occur predominantly within dense scrub. In California, by contrast, there is only scarce herb cover—twenty species making up just over 1 percent of the unburned slope vegetation—the herbs composed equally of annual and perennial species, all of which are native.

---

chaparral. A line drawn through the distance of 1.100, results in the following groupings:

a. narrow-leafed, open-growth form: A. fasciculatum *to* T. bicolor.

b. sclerophyllous, broad-leafed, evergreen shrub: C. greggii *to* Q. engelmannii.

c. mesophytic-leafed, drought-deciduous shrub: C. dumosa *to* R. trilobus.

d. spiny, drought-deciduous, chlorophyllous-stemmed shrub: T. trinervis *to* C. spinosa.

e. winter-deciduous tree or shrub: A. utahensis *to* N. obligqua.

f. leafless, chlorophyllous-stemmed shrub: E. andina *to* R. ephedra.

g. pungent-leafed, mesophytic shrub: A. californica *to* S. apiana.

h. weedy-habit shrub of varying form: E. fasciculatum *to* T. quinquinerva.

i. succulent rosette with spines: Y. schidigera *to* P. violaceae.

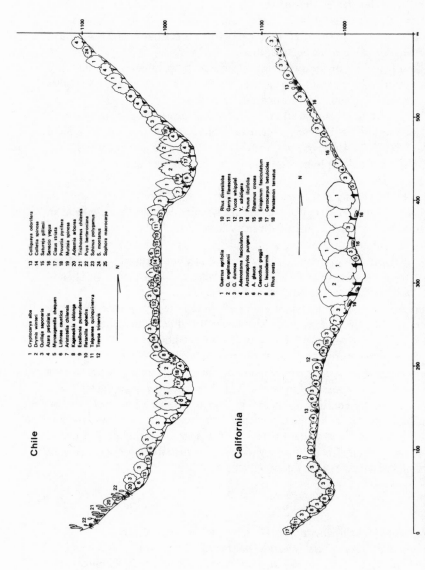

Chile

1 Cryptocarya alba	13 Colliguaya odorifera
2 Drimys winteri	14 Colletia spinosa
3 Quillaja saponaria	15 Satureja gilliesii
4 Azara petiolaris	16 Senecio yegua
5 Myrceugenella chequen	17 Cissus striata
6 Lithraea caustica	18 Proustia pyrifera
7 Aristotelia chilensis	19 Mutisia spinosa
8 Kageneckia oblonga	20 Adesmia arborea
9 Escallonia pulverulenta	21 Trichocereus chilensis
10 Retanilla ephedra	22 Puya berteroniana
11 Talguenea quinquinerva	23 Schinus polygamus
12 Trevoa trinervis	24 S. montanus
	25 Sophora macrocarpa

California

1 Quercus agrifolia	10 Rhus diversiloba
2 Q. engelmannii	11 Garrya flavescens
3 Q. dumosa	12 Yucca whipplei
4 Adenostoma fasciculatum	13 Y. schidigera
5 Arctostaphylos pungens	14 Prunus ilicifolia
6 A. glauca	15 Rhamnus crocea
7 Ceanothus greggii	16 Eriogonum fasciculatum
8 C. leucodermis	17 Cercocarpus betuloides
9 Rhus ovata	18 Penstemon ternatus

FIGURE 5-8. *Diagrammatic topographic and vegetation profile of the Chilean and Californian primary sites. (By E. Sierra Rafols.)*

100

TABLE 5-3 *Growth-form Diversity of the Primary-Site Shrub Communities*

Growth form	Number of species[a]	
	California	Chile
Broad-leafed evergreen	26	25
Broad-leafed deciduous	11	14
Narrow-leafed	6	4
Succulent	3	3
Spinose	0	6
Chlorophyllous-stemmed	0	2
Pinnate-leguminous	0	1

a. The species numbers given here exceed those in Table 5–4 because the range of habitats sampled was greater.
Source: Parsons (1973).

TABLE 5-4 *Percentage Shrub Cover by Slope at the Primary Sites in Chile and California*

Shrub species	Percent cover by slope orientation				Mean percent cover, all slopes
	North	South	East	West	
Chile: (Fundo Santa Laura)					
Cryptocarya alba	4.99	29.56	4.65	27.95	16.79
Colletia spinosa	0.15	21.73	10.58	0.14	8.15
Satureja gilliesii	3.74	3.52	12.79	11.41	7.87
Lithraea caustica	11.22	4.97	8.98	1.16	6.58
Trevoa trinervis	4.90	–	11.53	3.63	5.02
Colliguaya odorifera	7.03	0.08	2.47	7.26	4.21
Gochnatia fascicularis	0.59	4.00	5.21	2.72	3.13
Kageneckia oblonga	–	7.30	1.36	–	2.17
Baccharis linearis	1.68	2.41	2.20	0.69	1.75
Podanthus mitique	4.05	–	–	0.77	1.21
Teucrium bicolor	1.95	–	0.19	–	0.54
Retanilla ephedra	–	–	–	1.56	0.39
Quillaja saponaria	–	1.18	0.10	–	0.32
Haplopappus longipes	0.40	–	0.37	0.32	0.27
Muhlenbeckia hastulata	0.68	–	–	–	0.17
Puya berteroniana	0.67	–	–	–	0.17
Mutisia berteriana	–	–	–	0.63	0.16
Trichocereus chilensis	0.16	–	–	0.22	0.10
Ephedra andina	0.15	–	0.04	0.12	0.08
Ribes trilobum	–	0.22	0.06	–	0.07
Proustia pungens	–	0.06	–	0.04	0.06
Talguenea quinquinerva	0.23	–	–	–	0.06
Baccharis macraei	–	0.10	–	0.15	0.02
Total percent cover					59.29

TABLE 5-4 (Continued)

Shrub species	Percent cover by slope orientation				Mean percent cover, all slopes
	North	South	East	West	
California (Echo Valley)					
Adenostoma fasciculatum	56.37	122.22	82.10	146.33	101.86
Ceanothus greggii	58.65	2.51	56.77	80.82	49.69
Quercus dumosa	43.12	–	7.11	7.31	14.39
Arctostaphylos glauca	38.41	–	–	0.26	9.67
Eriogonum fasciculatum	2.30	9.58	–	9.43	5.33
Arctostaphylos glandulosa	0.50	–	16.85	0.50	4.46
Rhus ovata	6.33	0.73	1.45	–	2.13
Salvia clevelandii	–	2.09	6.11	–	2.05
Cercocarpus betuloides	7.08	–	–	–	1.77
Ceanothus leucodermis	3.23	–	–	–	0.81
Yucca whipplei	0.01	0.81	1.12	1.13	0.77
Quercus engelmannii	2.83	–	–	–	0.71
Rhamnus crocea	1.75	–	–	0.18	0.48
Trichostema parishii	–	0.52	0.72	0.14	0.35
Rhamnus ilicifolia	1.00	–	–	–	0.25
Salvia apiana	0.58	0.32	0.03	–	0.23
Penstemon ternatus	0.86	–	–	–	0.22
Artemisia californica	–	0.50	–	–	0.13
Lonicera subspicata	0.05	–	–	0.05	0.01
Total percent cover					195.31

Source: Keeley and Johnson (unpublished).

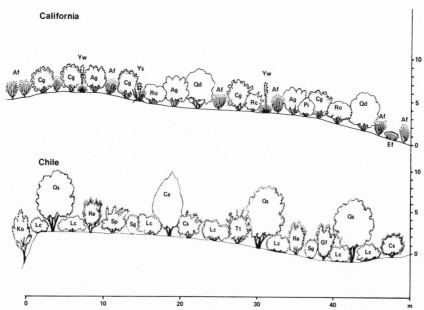

FIGURE 5-9. *Abstract profiles of the California chaparral (top) and the Chilean matorral. California: Af,* Adenostoma fasciculatum; *Ag,* Arctostaphylos glauca *and* A. glandulosa; *Cg,* Ceanothus greggii, *Ef,* Eriogonum fasciculatum; *Pi,* Prunus ilicifolia; *Qd,* Quercus dumosa; *Rc,* Rhamnus crocea; *Ro,* Rhus ovata; *Ys,* Yucca schidigera; *Yw,* Yucca whipplei. Chile: Ca, Cryptocarya alba; Cs, Colletia spinosa; Gf, Gochnatia fascicularis; Ko, Kageneckia oblonga; Lc, Lithraea caustica; Qs, Quillaja saponaria; Re, Retenilla ephedra; Sg, Satureja gillesii; Sp, Schinus polygamus; Tt, Trevoa trinervis. (By E. Sierra Rafols.)*

The short-term response of the vegetation to fire is somewhat dissimilar in California and Chile. Keeley and Johnson (1976) studied comparable one-year burns in both regions and found the establishment of shrubs by seedlings in Chile to be restricted to a single species, *Trevoa trinervis,* whereas, in California, they found post-fire seedlings of three species, *Ceanothus greggii, Arctostaphylos glauca,* and *Adenostoma fasciculatum.* They also found re-sprouting of shrubs common. In a general survey, Parsons (1973) noted that all of the Chilean shrub species he examined resprouted from root crowns after fire, whereas, in California, only about three-fourths of the shrub species resprouted after fire.

The post-burn differences in the herb floras were also striking. In California, there was a dramatic increase in the number of herb species (predominantly annuals) and in the amount of cover they constituted, when contrasted with the unburned vegetation (Table 5-5). In Chile, by contrast, fire was followed by a great reduction in number of herb species and in herb cover.

TABLE 5-5 *Herb Cover in Pre-Burn and One-Year Post-Burn Sites, at the Primary Sites in California and Chile*

	Percent cover	
Herb species	Pre-burn	Post-burn
Chile (Fundo Santa Laura), 1972		
Adesmia sp.	–	0.07
Adiantum glanduliferum	1.13	–
Alstroemeria neillii	0.06	–
Bowlesia uncinata	–	0.07
Calandrinia axilaris	–	0.07
Calceolaria glandulifera	0.06	–
Chlorea aurantiaca	0.13	–
Galium sp.	0.50	–
G. suffruticosum	–	0.07
Geranium dissectum	0.30	2.46
Gilia ramossissima	–	1.18
Gnaphalium purpureum	0.46	–
Madia sativa	1.20	0.07
Mutisia spinosa	2.06	–
M. subulata	0.20	–
Oxalis minima	–	0.07
Parietaria gracilis	–	1.19
Pasithaea coerulea	3.73	–
Piptochaetium panicoides	8.20	–
Quinchamalium gracilis	0.06	–
Sisyrinchium sp. and *S. pedunculatum*	1.83	1.73
Stachys grandidentata	0.20	–
Stellaria cuspidata and *Bowlesia tripartata*	4.66	–
Treptillion nodosa	0.13	–
Tropaeolum tricolor	0.06	–
Urtica urens	0.06	–
Valeriana magna and *V. crispa*	1.36	–
Native grasses: *Festuca magellenica, F. tertonensis, Melica violascens, Nasella chilensis, Trisetobromus hirtus*	1.72	–
Non-native grasses and herbs: *Briza minor, Bromus mollis, B. rigidus, Erodium botrys, E. cicutarium, Plantago lanceolata*	33.90	–
Total percent cover	62.02	6.98
California (Echo Valley), 1973		
Antirrhinum nuttallianum	–	0.92
Bromus rubens	–	0.02
Calochortus weedii	–	0.65
Calochortus sp.	0.22	–
Camissonia hirtella	–	1.15
Cerastium viscosum	0.14	–
Chaenactis artemisiifolia	–	26.70
Chorizanthe californica	–	0.77

TABLE 5-5 (Continued)

Herb species	Percent cover	
	Pre-burn	Post-burn
Cryptantha intermedia	0.38	9.57
Filago arizonica	–	0.02
Gilia caruifolia	–	8.97
Helianthemum scoparium	–	9.70
Lolium perenne	–	23.30
Lomatium lucidum	0.10	–
Lotus salsuginosus	–	22.00
L. scoparius	–	0.12
Mimulus brevipes	–	1.15
Nemacladus pinnatifidus	–	0.45
Phacelia brachyloba	–	6.97
Trichostema parishii	0.20	–
Total percent cover	1.04	112.46

Source: Data from Keeley and Johnson (1976), derived from 150-m line intercepts in
each condition and region.

Thus, the general contrast between regions finds Chile with a large herbaceous component in unburned vegetation and relatively little in post-burn
vegetation, whereas the opposite holds in California. Unfortunately, no
studies have been made of the long-term post-burn successional patterns in
Chile to contrast with the wealth of information on this vegetational aspect
of the southern California chaparral (see, for example, Hanes, 1971).

Summary of Vegetation Structure

A number of the vegetational features of the matched sites are remarkably similar. Most of these features center on the ecological characteristics of
the individual species, rather than on the structure of the vegetation itself.
For example, comparisons show that in each matched area the same shrub
types (in terms of such features as leaf characteristics, branching, and
sprouting behavior) occur, and that the matched sites of the two continents
are more similar in this respect than adjacent climatic sites in either continent.
Further, it was found that the habitat preference of a given shrub structural
form is generally the same on the two continents, a result that supports the
concept of convergence—in both functional and structural characteristics.

Although these vegetational analogies are pronounced at the individual
level, they become weaker at the community level; comparing climatically
matched sites, we find that in Chile, vis-à-vis California, the woody vegetation
constitutes less of the cover but more total species, and is characterized by

less dominance and greater height diversity. Detailed studies of the primary sites show that the Chilean site is more diverse than the Californian, both floristically and structurally. Tree forms are found in the valleys as well as on the mesic slopes in Chile, only in the valleys in California. In Chile, herbs are an important component of the shrub-dominated vegetation, but in California they become evident only after fire.

Possible explanations for these similarities and differences are developed in the following pages.

CARBON-GAINING STRATEGIES

The studies of the plant forms and vegetation types found in California and Chile demonstrate clearly that certain forms are characteristic of a given environmental regime. These forms have evolved independently in the two regions, to occupy the same distinctive habitat types. They are evidently adaptive for these habitats. But in order to determine more precisely how these growth forms are adaptive, we must relate to them the functions they perform. Since the most striking convergences between the plants of the two regions are in leaf features, it was obvious that carbon-gaining strategies are probably of considerable adaptive importance, and thus they became a focal point for study.

Geographic Trends and Principal Functional Types

A consideration of the principal plant types found within the span of climates from coastal arid to inland sites indicates a few basic forms: evergreen trees and shrubs, drought-deciduous shrubs, succulents, and drought-evading herbs. In both California and Chile the evergreen trees and shrubs predominate in the inland regions (Table 5-6), with the trees in the most mesic sites. In the drier coastal as well as semidesert regions, drought-deciduous shrubs increase in relative abundance. Succulents are quantitatively important only in the driest regions, and to an equivalent degree in both California and Chile. Drought-evading annuals are found abundantly throughout this aridity gradient.

Higher plants fix carbon by utilizing either the enzyme ribulose diphosphate carboxylase (C_3 plants) or phosphoenol pyruvate carboxylase (C_4 and crassulacean acid metabolism [CAM] plants) as their initial carboxylating enzyme. The CAM plants, which are all succulents, fix carbon during the night into organic acids; during the day, they close their stomata and refix carbon into sugars, utilizing light energy. This process is highly efficient in the relationship of carbon gained to water lost, although overall carbon-gaining efficiency is low. It is thus not surprising that they are most prevalent in arid regions (Mooney, 1972).

TABLE 5-6 *Characteristics of the Dominant Plants*

	Region and Locality					
	California			Chile		
Characteristic	Echo Valley	Camp Pendleton	San Telmo (Baja)	Fundo Santa Laura	Papudo	Cerro Potrerillo
Latitude	32°50′	33°15′	31°	33°10′	32°30′	30°25′
Climatic Type	mediterranean interior	mediterranean coastal	coastal desert	mediterranean interior	mediterranean coastal	coastal desert
Relative percent cover, by leaf type						
Evergreen	98.58	32.78	12.45	72.61	50.31	31.83
Drought-deciduous	1.41	67.08	62.70	11.55	36.95	41.76
Stem-chlorophyllous	0.00	0.00	0.00	14.54	0.00	0.37
Succulent	0.00	0.06	24.85	0.46	11.36	22.33
Unclassified	0.01	0.08	0.00	0.84	1.39	3.71
Relative percent cover, by photosynthetic type						
C_3	99.31	99.55	75.17	97.28	86.12	70.60
C_4	0.00	0.00	0.00	0.00	0.00	0.00
CAM	0.39	0.37	24.85	0.17	11.36	22.33
Unclassified	0.30	0.08	0.00	2.55	2.53	7.07

Source: Mooney, Troughton, and Berry (1974).

C_4 plants are also highly efficient in their ratio of carbon gained to water lost, because of their inherently high carbon-fixation capacities. These plants are distributionally centered in open tropical regions, where they evidently have recently evolved. Apparently, their greater water-use efficiency is of no competitive advantage in areas characterized by long summer droughts, such as those we studied, at least without additional adaptive features, for they are totally absent in the dominant floras (Table 5-6). They occur, rather, in neighboring climates where there is some rainfall during the hot period.

Excluding the succulents, then, the primary photosynthetic differentiation between the various plant types found along the mediterranean and semidesert aridity gradient is not biochemical but, rather, morphological and phenological. It will be useful to examine how these plants utilize such features in adapting to habitats of differing aridity.

The Evergreen Shrub and Tree

Evergreen shrubs photosynthesize the year round, although the rates of photosynthesis are considerably reduced if the summer drought is unusually severe (Dunn, 1970). An analysis of the environmental limitations on the photosynthetic capacity of a Californian evergreen sclerophyll has shown that water availability is the primary seasonal limitation on carbon gain, followed by photoperiod and, to a lesser degree, temperature (Figure 5-10) (Mooney et al. 1975). Thus, summer reductions in photosynthetic rate are due primarily to drought, and winter reductions are due to the short photoperiod. Toward the arid limits of distribution of the evergreen types, their photosynthetic capacity may be severely restricted by drought (Figure 5-11); and there they become competitively inferior to more drought-evading types.

The drought limitations of carbon gain in the evergreens are effected primarily through stomatal closure. With the replenishment of soil-moisture stores in the late fall, photosynthetic rates climb to pre-drought levels (Dunn, 1970).

There is little evidence that photosynthetic temperature acclimation by evergreen shrubs adjusts them to the seasonal thermal changes (Harrison, 1971). The effects of drought during the summer period apparently preempt any possible thermal metabolic adjustment.

Evergreen trees, such as *Arbutus menziesii* in California, are restricted to regions of higher latitude and higher rainfall. A comparison of the photosynthetic characteristics of *Arbutus* with the evergreen shrub *Heteromeles arbutifolia*, which is distributed in more arid low-latitude regions in California, showed that they have similar inherent photosynthetic and stomatal response characteristics (Morrow and Mooney, 1974). The shrub, however, has a deeper rooting system (as is characteristic of both the Californian and Chilean evergreen shrubs) and does not experience the comparable drought limitations of the tree (Figure 5-12).

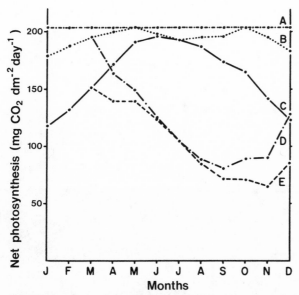

FIGURE 5-10. *Calculated seasonal course of photosynthesis for* Heteromeles arbutifolia, *a Californian evergreen sclerophyll, under conditions of: (A) no environmental limitations; (B) seasonal temperature limiting; (C) seasonal temperature and photoperiod limiting; (D) seasonal temperature and water limiting; (E) seasonal temperature, photoperiod, and water limiting (actual conditions) (Mooney et al., 1975).*

The Drought-Deciduous Shrub

The inherent photosynthetic capacity of mediterranean-climate drought-deciduous species averages twice that of evergreen types on a unit-leaf-area basis (Harrison et al., 1971) (Table 5-7). Further, these plants have higher rates of stomatal and cuticular water loss.

As the summer drought intensifies, the drought-deciduous plants gradually reduce their leafy evaporative surfaces. The Chilean drought-deciduous shrub *Proustia cinerea* was found to have essentially no stomatal control of water loss as the drought progressed, but, rather, gradually reduces its total leaf area. In contrast, the evergreen sclerophylls *Lithraea caustica* and *Kageneckia oblonga*, growing in the same area, control water during the drought by stomatal closure (Mooney and Kummerow, 1971).

Because of this drought-evading strategy, the deciduous species have a potential carbon-gaining period of only about six months, in contrast to the year-round capacity of the evergreens.

Although the drought-deciduous species bear leaves only for a limited

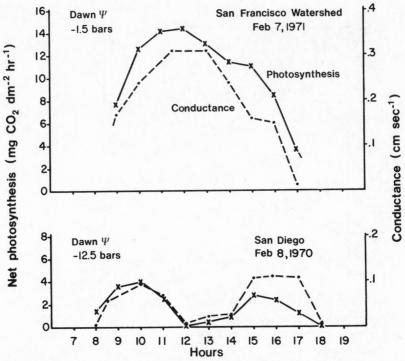

FIGURE 5-11. *Measured daily course of net photosynthesis and stomatal conductance of* Heteromeles arbutifolia *during late winter in northern California (San Francisco watershed) and at coastal San Diego, the latter near the arid limit of distribution of this species.* Ψ *indicates dawn xylem water potential. At the southern coastal habitat, this evergreen species is often under drought stress and exhibits depressed rates of carbon gain (Mooney et al., 1975).*

period, their high photosynthetic rates allow them to equal the carbon gain of the evergreens (Mooney and Dunn, 1970).

Summary of Carbon-Gaining Strategies

In California and Chile there are analagous arrays of adaptive carbon-gaining strategies, in which the predominance of each strategy corresponds to a position on an aridity gradient. At the moist, high-latitude stations, evergreen trees are found in both regions. With lower rainfall the shallow-rooted trees with large canopies confront summer water stress and give way to deep-rooted evergreen shrubs with smaller evaporative surfaces. As conditions become even drier at lower latitudes, the extended drought limits the carbon-gain period to such an extent that the evergreen canopy becomes a

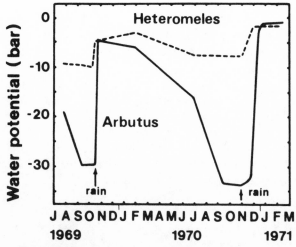

FIGURE 5-12. *Seasonal course of dawn water potential of an evergreen tree (*Arbutus menziesii*) and an evergreen shrub (*Heteromeles arbutifolia*) growing in a habitat where their distributions overlap (Morrow and Mooney, 1974). The shallow-rooted tree undergoes considerable drought stress each fall.*

TABLE 5-7 *Comparative Photosynthetic Rates of Mediterranean-Climate Plants* $(mg\ CO_2\ dm^{-2}\ hr^{-1})$

| | Maximum photosynthetic rate | |
| | | |
Growth forms	Observed on intact field plants	Specific example measured under optimal natural field conditions
Evergreen trees	–	*Arbutus menziesii* 15.0
		(San Francisco watershed)
Evergreen shrubs[a]	4.5 to 16	*Heteromeles arbutifolia* 13.9
		(San Francisco watershed)
Drought-deciduous shrubs[b]	15 to 42	*Salvia mellifera* 23.0
		(Mira Mar Mesa)
Succulents	–	*Dudleya farinosa*
		Day 5.4
		Night 1.6

a. Data from Dunn (1970) and Gigon (unpublished) and includes measurements on the Californian evergreens *Prunus ilicifolia, Rhus ovata, Heteromeles arbutifolia,* and the Chilean evergreens *Lithraea caustica, Kageneckia arbutifolia, Quillaja saponaria,* and *Colliguaya odorifera.*

b. Data from Harrison (unpublished) and Gigon (unpublished) and includes measurements on the Chilean *Podanthus mitique* and the Californian *Salvia mellifera, S. apiana, S. leucophylla,* and *Encelia californica.*

liability. In this region, drought-deciduous shrubs prevail, producing leaves only during the wet season; because of their mesophytic nature, they have a high capacity to fix carbon, and with the onset of the summer drought, they shed their leafy evaporative surfaces. At the driest point along the aridity gradient, succulent plants appear. They are the most economical of all plants in terms of ratios of carbon fixed to water lost. But because of their low growth rates they can persist only in the open desert habitats, where there is little competition for light.

That the same photosynthetic types appear in roughly the same proportions at comparable environmental positions in both California and Chile provides strong evidence that convergence has occurred, and that each adaptive type is optimal for a particular area on the aridity gradient.

PLANT SEASONALITY

An important functional activity of plants is the timing of carbon apportionment. At the gross level, this timing is manifested in such phenological events as stem and leaf initiation, leaf fall, flowering and fruiting. The timing of these events is primarily a response to selective forces within the environment. Thus, similarity of the phenological sequences in the plants of the matched climates of California and Chile would be evidence that these environments are indeed similar, and that the selective forces and adaptive responses have indeed been equal.

Our phenological studies were performed at several levels of detail ranging from simple seasonal observations of such events as flowering and fruiting, to determining the internal partitioning of carbon in selected species. Obviously, the simpler approaches could encompass a greater number of species.

We shall examine first the results of studies on a characteristic evergreen-sclerophyll shrub of the Californian chaparral, *Heteromeles arbutifolia*, as an example of the patterning of carbon allocation and its selective significance in mediterranean plants. More general comparisons will follow.

The Evergreen Type: Heteromeles as a Model

The evergreen-sclerophyll plants characteristic of the mediterranean regions fix carbon throughout the year, although at reduced levels during the summer and fall drought period and during the short winter days (Figure 5-13). Patterns of carbon use are, however, more markedly seasonal, and carbon outlay is rather precisely time-partitioned among many different plant functions.

The relationship between carbon gain and use over time was studied in detail for *Heteromeles arbutifolia*, which initiates its cambial activity in winter

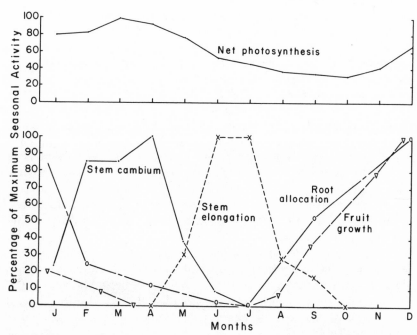

FIGURE 5-13. *Seasonal carbon gain and carbon allocation in the Californian evergreen sclerophyll* Heteromeles arbutifolia. *These data represent a synthesis from a number of studies and years: cambial growth, Avila et al. (1975), field plants at Echo Valley, 1971; stem elongation, Mooney, Parsons, and Kummerow (1974), garden plants at Stanford, California, 1970-71; fruiting biomass, unpublished data, field plants at Echo Valley, 1971; root allocation, Mooney and Chu (1974), potted plants growing in natural environment at Stanford, California, 1970-71.*

and continues it through spring (Figure 5-13). Stem elongation takes place rapidly in the spring, and is followed by an increase in the canopy biomass throughout the summer. The biomass of the fruit system develops during late summer and into the fall. Allocation of carbon to the roots also occurs during this period. Thus, at various times of the year photosynthetic carbon is shunted into diverse pathways.

Allocation patterns may differ with individual *Heteromeles* plants of different ages. For example, the period of stem elongation is longer in juvenile plants than in reproductive plants. This variance can be interpreted in terms of the overall carbon economy of the plant. In the juveniles, the "surplus" carbon that is not utilized in reproduction is shunted into canopy growth, and the increased growth capacity gives the young plants a greater competitive capacity (Mooney, Parsons, and Kummerow, 1974).

The seasonal flow of carbon in *Heteromeles* was examined at the chemical level, as well. It was found that the elaboration of storage compounds and of those secondary chemicals presumed to be effective in predator protection is greatest during the fall and winter. In spring and summer, carbon is shunted predominantly into compounds important in primary metabolism, and into cell-wall material (Figure 5-14) (Mooney and Chu, 1974).

Dement and Mooney (1974) found that the new leaves of *Heteromeles* contain relatively high levels of tannins, which increase to even higher amounts on maturation. The developing leaves also contain considerable amounts of nitrogen-bearing cyanogenic glucosides. As the leaves mature, the leaf-nitrogen level decreases, as does the quantity of cyanogenic glucoside. In the fall, after the initiation of the rainy season, cyanogenic-glucoside levels once again increase.

Leaf predation by herbivorous insects nonspecific to *Heteromeles* is greatest at the time of leaf initiation, when the leaves are the least sclerophytic and contain the least tannin.

In the young developing fruits of *Heteromeles* the pulp contains high levels of tannins and cyanogenic glucosides. As the fruits mature, these levels drop, whereas in the seeds the levels of cyanogenic glucosides increase. These relationships correlated with the low predation of developing fruits, and the bird dispersal of ripe fruits which contain protected embryos (Dement and Mooney, 1974).

Much of the information on the seasonality of apportionment in *Hetero-meles* has been synthesized by Mooney (1975) (Table 5-8). Although *Hetero-*

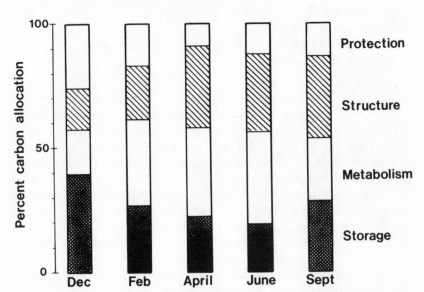

FIGURE 5-14. *Seasonal allocation of carbon to various functions in* Heteromeles arbutifolia *(Mooney and Chu, 1974).*

meles fixes carbon the year round, the carbon is apportioned nonuniformly to various plant parts and functions, presumably to best meet the demands of the internal as well as the external environment. It is presumed that canopy development puts priority demands on carbon in the spring, in response to competitive constraints on access to sunlight, leaving little carbon for root development, storage, or predator protection. In the fall and winter, however, when there is no canopy development, considerable carbon is shunted into these functions.

Flowering in *Heteromeles* generally occurs during early summer, after most other chaparral shrubs have flowered (Figure 5-15). This timing has been interpreted as a competitive response to the limited abundance of pollinators in the chaparral-shrub system (Mooney, Parsons, and Kummerow, 1974).

These studies indicate that many different selective forces determine the various phenological events in evergreens. It is thus unlikely that the timing of the various phenological events in a species is triggered by a single environmental cue. Further, the studies demonstrate that certain phenological events, such as canopy development, are likely to be synchronous in different plant species, especially in closed communities, whereas other events, such as flowering, are likely to be asynchronous.

TABLE 5-8 *Environmental Limitations on Carbon Apportionment and Their Physiological Consequences in* Heteromeles arbutifolia, *a Californian Sclerophyll Shrub*

Process	Environmental limitations	Physiological or evolutionary result
Carbon gain	Summer drought, winter cold, and short photoperiod	Rate reduction during winter and summer
Carbon allocation:		
Canopy	Competition for light, winter freeze, carbon limitations	Synchronous growth initiation in spring
Roots	Carbon limitation	Regulation of activity to noncanopy development period (winter-fall)
Flowers	Competition for pollinators	Flowering period asynchronous with other dominants
Protection	Carbon limitation	Regulation of activity to noncanopy development period (winter-fall)
Fruit dispersal	Germination conditions unsuitable during summer and fall	Winter fruit dispersal, with predator-protection mechanisms during long development period

Source: Modified from Mooney (1975).

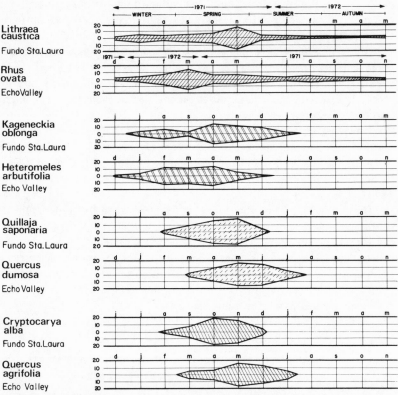

FIGURE 5-15. *Periods of canopy development, flowering, and fruiting for the dominant species in California during the spring of 1973, and in Chile during the 1972–73 growth period, at the primary sites. The slopes indicate the time course from zero to the plateau of new seasonal canopy-weight increase. Horizontal lines indicate duration of the flowering and fruiting periods, the latter data are for the year indicated, except where no activity was recorded, in which case the dates are for the year prior or following (Mooney and Kummerow, unpublished data).*

The Evergreen Type: Other Species

The more general seasonal aspects of the phenological development of a number of evergreen sclerophylls were studied in both California and Chile, with a concentration on species of the primary sites.

Avila et al. (1975) examined the seasonal cambial development of four Chilean and four Californian evergreens (Figure 5-16). Two of the species, both members of the Anacardiaceae, one in California and the other in Chile, exhibit year-round cambial activity, though both species show spring maxima. Two other shrub species studied, *Heteromeles arbutifolia* in California and

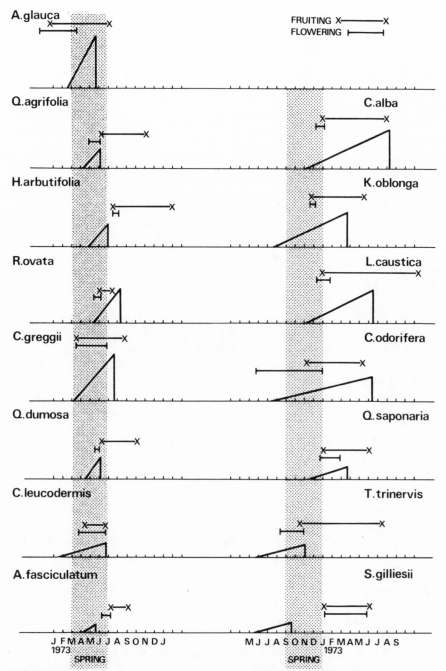

FIGURE 5-16. *Seasonal fluctuations in the cambial activity of matched Californian and Chilean evergreen sclerophylls (Avila et al., 1975). The sum of the numbers on each side of zero gives the percentage of the yearly xylem increase.*

Kageneckia oblonga in Chile, species that are remarkably similar on a morphological as well as distributional basis, exhibit similar cambial cycles: cambial activity is initiated in the winter and reaches a maximum in the spring.

The other four species studied were trees (except *Quercus dumosa*, which is a scrub-form variant of the evergreen oaks). The four pursue a very similar pattern of cambial activity, one that is restricted mostly to spring and early summer. The two Californian species start and end their activity somewhat later than their Chilean counterparts, a minor difference presumably related to the fact that the dry season starts about one month later in California than it does in Chile (di Castri, 1973).

Thus, all of the species, both Californian and Chilean, reach a spring peak of cambial activity, even though the species of Anacardiaceae in both California and Chile pursue some year-round activity. The tree species show a sharper activity peak than the shrub species.

We also compared the canopy-growth period (i.e., the seasonal increase in terminal leaves and stems) of a group of the dominant Chilean and Californian species (Figure 5-15). The major canopy-growth period of the Californian species is restricted to the spring. The growth period of the Chilean species is much more protracted, extending from winter for certain species, such as *Trevoa trinervis* and *Satureja gilliesii*, through summer, for others, such as *Lithraea caustica*. *Trevoa* is drought-deciduous, *Satureja* nearly so; the Californian species are all evergreens. These canopy-growth patterns contrast with the anomalous, predominantly summer growth rhythms characteristic of shrubs of the Australian mediterranean climatic region (Specht, 1973).

The flowering period of the dominant evergreens is also more protracted in Chile than in California, extending from winter through fall (Figure 5-15). These cross-continental differences become even more pronounced when we consider more of the species in each community, as we shall see (p. 123).

Even among the various Californian species, which on the whole exhibit a fairly synchronized canopy-growth period, the patterns of reproductive activity are dissimilar (Figure 5-15). Certain species, such as *Arctostaphylos glauca*, flower and fruit prior to maximum canopy growth; others, such as *Ceanothus leucodermis*, reproduce during peak growth; and others, such as *Heteromeles arbutifolia*, reproduce after most of the canopy growth ceases. These same general phenological types are present in Chile. Furthermore, in both California and Chile there are species that have a brief flowering period followed by a long fruit-development period; these include the *Quercus* spp. and *Heteromeles* in California and *Cryptocarya, Trevoa, Lithraea*, and *Kageneckia* in Chile. Finally, there are a number of species that have a long flowering period during which there is continuous initiation and development of fruit.

These diverse reproductive strategies mean that some species reproduce on the gains of the preceding year, whereas others do so on the current year's photosynthate. The marked year-to-year variability of the climate, and hence photosynthetic gains, has a profound influence on the community response.

During 1972, the Echo Valley site was particularly dry (62 percent of normal precipitation at nearby Descanso). There was virtually no growth in the dominant plants that year (Figure 5-17). Certain species were extremely sensitive to this drought and undertook no reproductive development at all (e.g., *Heteromeles arbutifolia* (Figure 5-18). This species, as indicated earlier, presumably produces reproductive tissue on current photosynthate subsequent to canopy development. In contrast, *Ceanothus leucodermis* (Figure 5-19), which apparently produces reproductive tissue in part from storage reserves, is not as directly attuned to current environmental conditions. *Ceanothus* flowered during the drought year but did not set fruit. Furthermore, during the year following the drought there was no reproductive activity at all in *Ceanothus*, indicating reserve depletion. This dependence of reproductive activity on current-year photosynthate in some species, and on reserves from the previous year in others, means that, barring a long-term drought, at least one of the dominants will go into the reproductive mode in any given year.

The Drought-Deciduous Type

Less detailed information is available on the developmental sequences in plants of the drought-deciduous type. In Chile, drought-deciduous types are

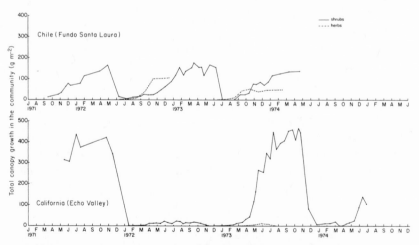

FIGURE 5-17. *Seasonal change in standing biomass of shrubs (solid lines, 91% of the relative cover in California and 72.5% in Chile accounted for) and total herbs (dashed lines) at the Chilean and Californian primary sites. Note that during the dry year 1972 in California there was virtually no production (Mooney and Kummerow, unpublished data). Herb biomass measured only during the years indicated.*

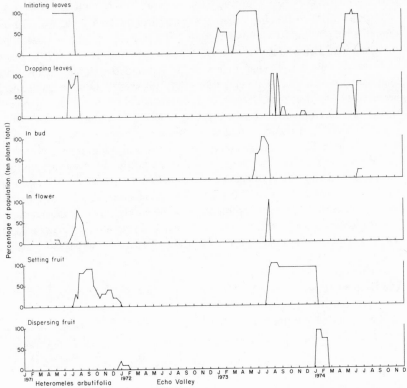

FIGURE 5-18. *Seasonal development of* Heteromeles arbutifolia *at the Californian primary site. The figures given are the percentage of ten plants at a given state. No observations on bud made in 1971 (Mooney and Kummerow, unpublished data).*

represented within the predominantly evergreen vegetation at the primary site, as noted earlier (*Trevoa trinervis* and *Satureja* in part). These species are the first to develop their canopy (Figure 5-15; see also Mooney and Kummerow, 1971). In California also, wherever drought-deciduous and evergreen types co-occur, this sequence holds. The drought-deciduous types evidently utilize the water in the uppermost soil layers for their growth. Even in particularly bad drought years, they may grow, though the evergreen species will not (Harvey and Mooney, 1964).

Aljaro et al. (1972) noted that the cambial activity of the drought-deciduous species *Proustia cuneifolia*, which is found at the Chilean primary site, starts soon after the winter rains commence and ceases with the end of the rainy period about the end of October. Thus, its period of cambial activity is generally much earlier than that reported for most evergreens growing in the same area (Avila et al., 1975). This species, as noted, sheds its leaves during the drought period. It flowers and fruits subsequent to leaf fall, hence on the

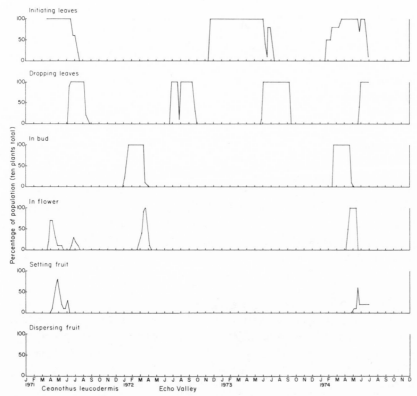

FIGURE 5-19. *Seasonal development of* Ceanothus leucodermis *at the Californian primary site. The figures given are the percentage of ten plants at a given state. No observations on bud made in 1971 (Mooney and Kummerow, unpublished data).*

strength of reserves accumulated during that growing season. Canopy production during the subsequent season must also be built in part on these same reserves. A similar pattern has been noted for the Californian drought-deciduous tree species *Aesculus californica* and *A. parryi* (Mooney and Hays, 1973; Mooney and Bartholomew, 1974).

In a comparison of the carbohydrate economy of co-occuring Californian trees, one drought-deciduous and the other evergreen, Mooney and Hays (1973) concluded that the evergreen has a more buffered economy; i.e., there are no dramatic storage-utilization periods in the evergreens, as there are in the drought-deciduous species. Furthermore, reproductive growth draws at least in part upon current photosynthate in the evergreen species, rather than entirely from reserves. The developmental response of the two types may in one case reflect part climatic events, and in the other, current events. But there is variability within the evergreens themselves in this respect, as we have seen.

We conclude, then, that the drought-deciduous plants are more opportunistic than the evergreen species. They develop earlier than the evergreens, usually when the first rains come, and they do so apparently at the expense of reserves. Their generally shallower rooting system is an essential requisite for this pattern of earlier exploitation of the available water (Harvey and Mooney, 1964, Mooney and Dunn, 1970).

Annual Species

The annual herbs represent a phenological type that is even more opportunistic than the drought-deciduous species. They become active (germinate) with the first rains and reach their peak production in early spring (Figure 5-17; the Californian herb production within the chaparral is so low that the early growth phases are not evident in this figure). More extensive growth data from the Californian and Chilean grasslands, both of which support predominantly annual floras, show comparable seasonalities (Gulmon, unpubl. data).

The initiation of growth in a given annual varies greatly from one year to another, since it is closely dependent on rainfall initiation, which itself is highly variable between years. Reproduction, however, is more regular between years for any given species (Slade et al., 1975). As with shrubs, species show greater similarity in canopy-growth period than in the timing of reproduction.

Phenology of Community Groups

At each of the two primary sites, we studied the phenological development of about eighteen species. These observations were confined to the perennials, but included virtually all of the growth types present: drought-deciduous shrubs and subshrubs, evergreen trees and shrubs, and succulents.

A summary of these observations reveals a distinct difference between primary sites. Virtually all phenological events are more protracted within the Chilean community than within the Californian (Figure 5-20). These community differences can be attributed in part to the differences in growth-form diversity between areas. As noted above, the phenological development of the various growth forms differs, with the earliest development in the annuals, followed by the drought-deciduous species and, finally, the evergreens. In Chile, there is a greater growth-form diversity within the community, and hence a protraction of events in Chile. Further, there are small climatic differences between the two regions which may be responsible for the dissimilarity in growth patterns.

Common garden observations on a set of woody species originating in the primary-site regions of California and Chile indicate a much greater sensi-

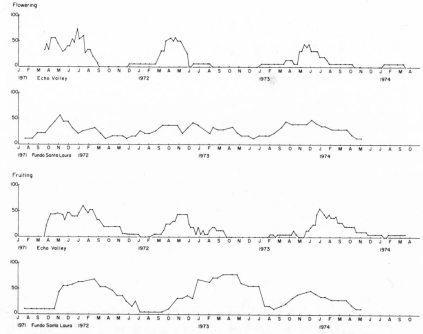

FIGURE 5-20. *Percentage of the dominant tree, shrub, and subshrub species flowering and fruiting at various times of the year at the primary sites of California and Chile. Data are for 18 species in California (Echo Valley) and 20 species in Chile (Fundo Santa Laura). Other events such as leaf initiation, flower bud, fruiting, fruit dispersing, and leaves falling show the same differential between sites, i.e., less seasonality in Chile (Mooney and Kummerow, unpublished data).*

tivity to frost in the Chilean species than in the Californian species (Mooney, unpubl. data).

Summary of Plant Seasonality

In both the Californian and Chilean communities, there is a partitioning of resources over time. The partitioning schedule is correlated to growth form: the herbs, then the drought-deciduous shrubs, and, finally, the evergreen shrubs, progressively initiate growth activity following the breaking of the drought by the winter rains. Within growth forms, however, there is generally a strong synchronization of vegetative growth, although reproductive activity may be asynchronous. This generalization is particularly valid for the Californian evergreen shrubs, all of which have a strong spring-growth period. In Chile, some evergreen species show a summer-growth rhythm.

Even though carbon gain proceeds continuously throughout the year in the evergreen shrubs, carbon use is strongly time-partitioned to diverse growth activities (canopy growth, root growth, reproductive growth) in a competitively adaptive manner.

Among the shrub species of both communities, some members produce flowers and fruits on stored photosynthate, others on current photosynthate. This results in a variable reproductive output between species from year to year, some output patterns being closely linked with current weather, others with past weather. The result is a more buffered output of reproductive carbon into the community than would be predicted from current productivity. As we shall see, this is also true for leaf-litter fall in these communities.

The phenological development of the plants of the Chilean evergreen community is more protracted than that of its Californian counterpart, owing in part to the greater diversity of growth forms in the Chilean vegetation, but also to the small climatic differences between the two areas.

BIOMASS DISTRIBUTION AND NET PRODUCTIVITY

In the preceding sections we examined the morphological characteristics of analogous Chilean and Californian plants and related these characteristics to certain physiological attributes of the plants, such as carbon-gaining capacity and the timing of carbon apportionment. Here we make a different link between morphology and function, one that better integrates environmental influences through time. In this instance we shall examine the distribution and configuration of plant biomass in the community, and study certain changes in these characteristics with time.

Biomass Distribution

The above-ground portions of representatives of the dominant woody species in both California and Chile were harvested at the primary sites. With the exception of *Trevoa trinervis* in Chile, the plants selected were all evergreens. We also made comparative harvests of dominant members of the more arid coastal drought-deciduous community in California (Table 5-9).

In evergreen communities of both primary sites, the shrubs covered an average of 1.5 m^2 and were less than 2 m tall. The mean total leaf weights of the shrubs at the two sites were comparable, although the overall plant weight of the Chilean shrubs was smaller. The shrubs of the drought-deciduous community were somewhat smaller in overall size, and considerably smaller in overall biomass, than those of either of the evergreen communities.

On a weight-per-area basis, the Californian shrubs of the evergreen community were the heavier, averaging nearly 3 kg m^{-2} (Table 5-10). The Chilean

TABLE 5-9 *Mean Plant Dimensions of the Dominants of the Evergreen Communities at the Californian and Chilean Primary Sites, and of a Californian Drought-Deciduous Community*[a]

Species		Height (m)	Diameter (m)	Projected areal cover (m²)	Total shoot weight (g)	Total leaf weight (g)
California						
Evergreen community						
Rhus ovata		1.48	1.81	2.58	5,462.0	1,187.9
Ceanothus leucodermis		1.75	1.11	0.96	2,077.9	282.7
Heteromeles arbutifolia		1.68	1.56	1.91	6,270.8	1,220.0
Arctostaphylos glauca		1.76	1.36	1.44	4,839.0	1,346.1
Adenostoma fasciculatum		1.16	0.98	0.76	1,311.9	221.0
Quercus agrifolia[b]		2.08	1.49	1.74	8,015.8	805.1
Q. dumosa		1.52	1.37	1.47	3,007.0	514.4
Ceanothus greggii		1.96	1.24	1.21	4,092.8	720.6
	Mean	1.67	1.35	1.51	4,384.7	787.2
Drought-deciduous community						
Encelia californica		0.90	0.79	0.49	120.6	22.2
Salvia mellifera		1.35	1.30	1.33	1,174.8	306.7
Artemisia californica		1.10	1.24	1.21	1,112.4	77.5
	Mean	1.12	1.11	1.01	802.6	135.5
Chile						
Evergreen community						
Lithraea caustica		0.80	2.09	3.43	2,966.9	1,011.8
Trevoa trinervis		1.40	1.90	2.85	2,355.5	446.3
Kageneckia oblonga		1.24	1.02	0.81	1,085.9	400.1
Colliguaya odorifera		1.12	1.94	2.94	4,072.8	736.5
Satureja gilliesii		0.92	0.82	0.52	399.4	59.3
Cryptocarya alba[b]		1.44	1.51	1.78	3,017.7	1,360.3
Quillaja saponaria[b]		1.52	1.81	2.57	3,825.7	929.7
	Mean	1.21	1.58	1.76	2,532.0	706.3

a. Based on mean values for five mature specimens per species. Detailed sampling information for the data given in Tables 5-9 to 5-17 will be published elsewhere (Mooney et al.).
b. Shrub forms.

evergreen shrubs were substantially lighter, yet heavier than the drought-deciduous shrubs of California.

The percentage of dry matter taken up by leaf material was near 20 percent for the shrubs of all communities. This is a remarkably high percentage of dry leafy material for plant communities. For example, deciduous forests generally have less than 2 percent by dry weight of above-ground biomass in leaves, and evergreen coniferous forests somewhat over 10 percent (Rodin and Basilevic, 1968). A mixed coniferous and deciduous forest described by Whittaker and Woodwell (1969) had a leaf-biomass percentage of 6.7.

The average unit leaf area per unit ground surface (leaf area index, LAI) was 2 to 2.6 for the plants of the evergreen community, and only 1.3 for the drought-deciduous shrubs. We found a broad variance among the different shrub species, the LAI ranging from 1 to 4. LAI values calculated for the communities as a whole are similar to the mean shrub values given in Table 5-10, although the Chilean community values were somewhat lower owing to the low total plant cover (LAI = 1.6) in Chile.

TABLE 5-10 *Biomass Distribution of the Dominant Plants of the Evergreen Communities at the Primary Sites, and of a Californian Drought-Deciduous Community*[a]

Species	Biomass distribution, above ground			Percent stems	Percent leaves	Leaf-area Index (m² m⁻²)	Bark-area Index (m² m⁻²)
	Stems (g m⁻²)	Leaves (g m⁻²)	Total shoot (g m⁻²)				
California							
Evergreen community							
Rhus ovata	1,656.7	460.4	2,117.1	78.3	21.7	1.95	0.95
Ceanothus leucodermis	1,870.0	294.5	2,164.5	86.4	13.6	2.11	0.87
Heteromeles arbutifolia	2,644.6	638.7	3,283.3	80.5	19.5	2.81	1.39
Arctostaphylos glauca	2,425.6	934.8	3,360.4	72.2	27.8	3.58	1.17
Adenostoma fasciculatum	1,435.4	290.8	1,726.2	83.2	16.8	3.09	2.00
Quercus agrifolia[b]	4,144.1	462.7	4,606.8	90.0	10.0	3.61	1.88
Q. dumosa	1,695.7	349.9	2,045.6	82.9	17.1	2.46	1.17
Ceanothus greggii	2,787.0	595.5	3,382.5	82.4	17.6	1.61	1.86
Mean	2,332.4	503.4	2,835.8	82.0	18.0	2.65	1.41
Drought-deciduous community							
Encelia californica	200.8	45.3	246.1	81.6	18.4	0.73	0.73
Salvia mellifera	652.7	230.6	883.3	73.9	26.1	2.43	2.39
Artemisia californica	855.3	64.0	919.3	93.0	7.0	0.78	2.40
Mean	569.6	113.3	682.9	82.8	17.2	1.31	1.84
Chile							
Evergreen community							
Lithraea caustica	570.0	295.0	865.0	65.9	34.1	1.14	—
Trevoa trinervis	669.9	156.6	826.5	81.0	19.0	2.26	—
Kageneckia oblonga	846.6	494.0	1,340.6	63.2	36.8	2.24	—
Colliguaya odorifera	1,134.8	250.5	1,385.3	81.9	18.1	0.91	—
Satureja gilliesii	654.1	114.0	768.1	85.2	14.8	1.39	—
Cryptocarya alba[b]	931.1	764.2	1,695.3	54.9	45.1	4.42	—
Quillaja saponaria[b]	916.7	361.8	1,278.5	71.7	28.3	1.75	—
Mean	817.6	348.0	1,165.6	72.0	28.0	2.02	—

a. The values given are for the area occupied by the indicated species and not for the community as a whole. In order to obtain community values the percent cover of the species must be considered.
b. Shrub forms.

Community LAI values of 2 also characterize the evergreen-sclerophyll communities of Australia (heathland) (Jones, 1968). In contrast, temperate-forest communities have LAI values of generally about 5 (3–8) (Carlisle et al., 1966; Whittaker and Woodwell, 1969). Evergreen-shrub communities in mesic climates have an LAI of over 4 (mixed heath balds of the Great Smoky Mountains; Whittaker, 1966).

The features of the leaves of the plant dominants of the two evergreen communities have remarkably similar average values, despite a degree of variability in certain characteristics, such as absolute size and weight (Table 5–11). The average values for the dominant plants for *both* communities on the two continents lie within narrow limits: size, 3.7 to 4.3 cm^2; weight, 81.4 to 90.5 mg; specific leaf weight, 18.7 to 19 mg cm^{-2}; percent lignin, 9.3 to 9.8; percent cellulose, 11.2 to 11.8, percent fiber, 21 to 21.1; and sclerophyll index (percent fibers $\times$ 0.64 [= percent crude fiber] $\times$ 100 + nitrogen $\times$ 6.25 [= crude protein]) (Loveless, 1962), 241.5 to 290.5.

The sclerophyll-index values are generally very high, and several exceed the upper limit of the values given by Loveless (1962) for leaves from a wide range of habitats. It is noteworthy also that in each community there is a species with a particularly low sclerophyll index; these are the analog species *Ceanothus leucodermis* and *Trevoa trinervis*, both of the Rhamnaceae and both possibly nitrogen fixers and somewhat early-successional.

The leaf sizes and specific weights of the dominant plants of the Chilean and Californian evergreen communities may be contrasted with those of the evergreen shrubs of the Great Smoky Mountains studied by Whittaker (1962). The leaves of the latter were generally much larger, but had a specific weight averaging somewhat less than one-half that of the mediterranean shrubs.

The average above-ground biomass values for the various shrub species have been utilized to calculate the biomass of the community as a whole (Table 5–12). The percentage cover constituted by each of the various species within the community was multiplied by its biomass-per-area figure and corrected for the percentage of the community that that species represented. Performing these calculations, we find that the biomass for the California community is 2,308 g m^{-2}; that for Chile, 738 g m^{-2}. Thus, the figures differ between communities by a factor of 3. On a communitywide basis, the biomass values are reduced from the shrub average somewhat, in California because of the dominance of *Adenostoma*, a relatively light shrub, and in Chile because of the low total plant cover.

Net Production

At bimonthly intervals during the year, the terminal shoots of the current year were harvested from the dominant species of both the Californian and Chilean evergreen communities. From these harvests, estimates of the course of yearly production of shoot components were made. The values given in

TABLE 5-11 *Mean Leaf Characteristics of the Dominant Plants of the Evergreen Communities at the Californian and Chilean Primary Sites*

Species	Leaf size (cm²)	Leaf weight (mg)	Specific leaf weight (mg cm⁻²)	Percent lignin[a]	Percent cellulose[a]	Percent fiber[a] (lignin & cellulose)	Percent nitrogen	Sclerophyll index	Percent phosphorus
California									
Rhus ovata	13.0	273.0	21	11.1	8.6	19.7	0.71	284.1	.05
Ceanothus leucodermis	1.4	16.8	12	4.6	7.0	11.6	1.61	73.8	.09
Heteromeles arbutifolia	7.2	165.6	23	8.6	9.3	17.9	0.75	244.4	.04
Arctostaphylos glauca	5.2	140.4	27	12.0	9.2	21.2	0.58	374.3	.04
Adenostoma fasciculatum	0.1	0.8	8	10.4	9.2	19.6	0.65	308.8	.04
Quercus agrifolia[b]	4.4	57.2	13	10.5	20.4	30.9	1.15	275.1	.08
Q. dumosa	2.1	29.4	14	11.7	18.2	29.9	1.30	235.5	.08
Ceanothus greggii	1.1	40.7	37	9.5	8.0	17.5	1.29	138.9	.05
Mean	4.3	90.5	19.4	9.8	11.2	21.0	1.01	241.9	.06
Chile									
Lithraea caustica	5.8	150.8	26	18.8	13.9	32.7	0.65	515.2	.08
Trevoa trinervis	1.1	8.8	8	2.2	6.2	8.4	1.66	51.8	.08
Kageneckia oblonga	7.3	153.3	21	8.0	12.2	20.2	0.82	252.3	.12
Colliguaya odorifera	2.0	54.0	27	10.0	8.6	18.6	0.77	247.4	.06
Satureja gilliesii	0.1	0.8	8	4.9	18.2	23.1	0.76	311.2	.13
Cryptocarya alba[b]	5.9	118.0	20	10.2	15.3	25.5	0.73	357.7	.07
Quillaja saponaria[b]	4.0	84.0	21	11.0	8.2	19.2	0.66	297.9	.05
Mean	3.7	81.4	18.7	9.3	11.8	21.1	0.86	290.5	.08

a. Unpublished data of Cromack.
b. Shrub forms.

Table 5-13 are the mean weights of the various components measured on fully developed material, and thus represent the total leafy twig production for the year.

The average total production per unit area was over one-third greater in the Californian shrubs than in the Chilean shrubs. The distribution of production, however, was quite comparable: at both sites, average production was approximately 70 percent for leaves; 17-22 percent for stems (including rachis); and 8-11 percent for reproductive parts. Production was fairly constant during three years of study in Chile, but fluctuated greatly in California (Table 5-14). In California, during the drought year of 1972, production was only about 15 percent of that during the high-rainfall year of 1973.

Calculations of productivity for the high-production year of 1973 in California and for the 1972-73 season in Chile give values of 671 g m^{-2} yr^{-1} and 253 g m^{-2} yr^{-1}, respectively (Table 5-12). Although the productivity of the Californian community is thus over twice that of the Chilean community, the ratios of shrub biomass to shrub production are quite comparable, 3.4 for California and 2.9 for Chile. The differences in productivity between the two communities thus reflect not a fundamental disparity but rather the difference in biomass between the two.

A portion of this community difference in productivity is made up by the greater herb production in Chile (Table 5-14); the lower shrub cover in Chile results in greater herb development.

The values of shrub productivity based on terminal-twig growth do not, of course, reflect total productivity, since they account neither for older stem and leaf growth nor for root growth. According to Whittaker (1962), total annual shoot production of shrubs is about twice the value of terminal twig growth alone. Utilizing this finding, we can very roughly calculate that net aboveground production would be in order of 1,340 g m^{-2} yr^{-1}. Again, according to Whittaker's conversion figures for the Great Smoky Mountain shrubs, consideration of root production would double this figure to 2,640 g m^{-2} yr^{-1}. Although these values are rough approximations, they are in the same range as calculations based on the gas-exchange measurements of Dunn (1970; see Table 5-15). Furthermore, they fall in the high range of values for temperate evergreen forests cited by Kira and Shidei (1967). Our values may be overestimates. Above-ground annual increment of chaparral biomass based on standing crop divided by age give values of only about 125 gm^{-2}yr^{-1} (Specht, 1969). Addition of annual litter production of about 270 gm^{-2}yr^{-1} to this would still be considerably less than the values given above.

Litter Production

Annual production of litter in the evergreen community in California varied from year to year (Table 5-14), but not to the same degree as did terminal-twig production, owing to the fact that the leaves on the shrubs gener-

TABLE 5-12 *Standing Biomass, Net Productivity and Litter Production of the Dominant Plants of the Evergreen Communities of the Californian and Chilean Primary Sites*

Species	Percent cover in community[a]	Above-ground biomass (g m^{-2})	Relative within community (g m^{-2})	Above-ground twig production (g m^{-2} yr^{-1}) 1973
California				
Rhus ovata	1.1	2,117.1	23.3	1,027.8
Ceanothus leucodermis	0.4	2,164.5	8.7	204.6
Heteromeles arbutifolia	–	3,283.3	–	667.5
Arctostaphylos glauca	5.0	3,360.4	168.0	1,394.6
Adenostoma fasciculatum	52.2	1,746.2	901.1	541.0
Quercus agrifolia[b]	–	4,606.8	–	338.5
Q. dumosa	7.4	2,045.6	151.4	423.3
Ceanothus greggii	25.4	3,382.5	859.2	859.9
Total of above	91.5		2,111.7	
Total for community	100.0		2,307.9	
Chile				1972–73[c]
Lithraea caustica	6.6	865.0	57.1	504.2
Trevoa trinervis	5.0	826.5	41.3	383.8
Kageneckia oblonga	2.2	1,340.6	29.5	643.0
Colliguaya odorifera	4.2	1,385.3	58.2	438.2
Satureja gilleisii	7.9	768.1	60.7	208.4
Cryptocarya alba	16.8	1,695.3	284.8	483.6
Quillaja saponaria	0.3	1,278.5	3.8	180.1
Total of above	43.0		535.4	
Total for community	59.3		738.4	

a. Cover values are normalized to 100% for California, since overlapping of plants did not result in an increase in biomass per unit area.
b. Shrub forms.
c. Growth period in Chile spans months in both calendar years.

ally persist more than a year. Further, litter production (leaf drop) can be increased during drought years when there is little new leaf production. Although there was considerable variation in litter production between years for individual Chilean shrub species, the yearly mean values for all species were similar.

In California as well as in Chile, the bulk of the litter was composed of leaves, in all species (except *Quercus agrifolia*, for which there was also a large percentage of bark) (Table 5-16). With respect to the other species in California, which were all generally even-aged at twenty-three years, the oaks were presumably quite old, since they had not been destroyed in the most recent fire. The bark fraction of litter in evergreen forests has been seen to increase with the age of the stand (Ando, 1970).

Relative within community (g m⁻² yr⁻¹)	Biomass-production ratio	Litter (g m⁻² yr⁻¹) 1973	Relative within community (g m⁻² yr⁻¹)	Leaf production (g m⁻² yr⁻¹)	Relative within community (g m⁻² yr⁻¹)
11.3		380.3	4.2	703.3	7.78
0.8		314.6	1.3	142.3	0.57
–		254.8	–	395.3	–
69.7		531.3	26.6	1,146.9	57.35
282.4		213.2	111.3	126.4	65.98
–		646.0	–	296.6	–
31.3		117.3	8.7	369.9	27.37
218.4		383.0	97.3	601.4	152.76
613.9			249.4		311.81
670.9	3.4		272.6		340.78
		1972–73[c]			
33.2		290.3	19.2	374.3	24.70
19.2		176.2	8.8	166.8	8.34
14.1		209.9	4.6	453.6	9.58
18.4		427.5	18.0	278.2	11.68
16.5		155.6	12.3	135.3	10.69
81.2		326.1	54.8	417.6	70.16
0.5		132.2	0.4	158.0	0.47
183.1			118.1		135.62
252.5	2.9		162.9		187.03

Our estimate of litter production in the Californian community during 1973 is 273 g m⁻² yr⁻¹ (Table 5-12). This value is almost identical to the average value cited for litter fall (280 g m⁻² hr⁻¹) for a chaparral stand in nearby Los Angeles County for a number of years (Kittredge, 1955). Litter fall in the Chilean community, as well as that for an Australian heathland (Jones, 1968), is about one-half of this value.

Litter fall for many species in both Chile and California peaks soon after the initiation of the new canopy (Figures 5-21 and 5-22). Certain species, however, such as the oaks (*Quercus* spp.) in California and *Satureja* in Chile, have fairly uniform litter production throughout the year.

The production of new leaves (Table 5-13) is in most cases different from, usually greater than, the production of leaf litter, for a number of reasons. First, leaf production is not synchronized with leaf-litter production,

TABLE 5-13 *Terminal Shoot Production at the Californian and Chilean Primary Sites*

Species	Terminal shoot production (g m^{-2})							Percentage of total					
	Leaves	Stems	Flowers	Fruits	Rachis	Buds	Total	Leaves	Stems	Flowers	Fruits	Rachis	Buds
California (1973)													
Rhus ovata	707.3	135.6	8.4	0	0	176.5	1,027.8	68.8	13.2	0.8	0	0	17.1
Ceanothus leucodermis	142.3	62.3	0	0	0	0	204.6	69.6	30.4	0	0	0	0
Heteromeles arbutifolia	395.3	49.5	77.8	120.8	24.1	0	667.5	59.2	7.4	11.7	18.1	3.6	0
Arctostaphylos glauca	1,146.9	177.4	0	9.0	70.3	0	1,394.6	82.2	12.7	0	0	5.0	0
Adenostoma fasciculatum	126.4	65.1	158.6	130.0	60.9	0	541.0	23.4	12.0	29.3	24.0	11.3	0
Quercus agrifolia[a]	296.6	25.3	2.1	14.5	0	0	338.5	87.6	7.5	0.6	4.3	0	0
Q. dumosa	369.9	47.9	0.7	4.8	0	0	423.3	87.4	11.3	0.2	1.1	0	0
Ceanothus greggii	601.4	204.1	0.0	0	0	54.4	859.9	69.9	23.7	0	0	0	6.3
Mean	473.3	95.9	31.0	33.8	19.4	28.9	682.2	68.4	14.8	5.3	5.9	2.5	2.9
Chile (1972–73)													
Lithraea caustica	374.3	67.9	7.0	55.0	—	—	504.2	74.2	13.5	1.4	10.9	—	—
Trevoa trinervis	166.8	185.5	14.3	17.2	—	—	383.8	43.5	48.3	3.7	4.5	—	—
Kageneckia oblonga	453.6	112.7	31.2	45.5	—	—	643.0	70.5	17.5	4.9	7.1	—	—
Colliguaya odorifera	278.2	84.4	19.2	56.4	—	—	438.2	63.5	19.2	4.4	12.9	—	—
Satureja gilliesii	135.3	61.0	2.5	9.6	—	—	208.4	64.9	29.3	1.2	4.6	—	—
Cryptocarya alba[a]	417.6	66.0	0	0	—	—	483.6	86.4	13.6	0	0	—	—
Quillaja saponaria[a]	158.0	21.1	1.0	0	—	—	180.1	87.7	11.7	0.6	0	—	—
Mean	283.4	85.5	10.7	26.2	—	—	405.9	70.1	21.9	2.3	5.7	—	—

a. Shrub forms.

TABLE 5-14 *Maximum Observed Current Terminal Shoot Weights and Litter Production of the Dominant Plants of the Evergreen Communities at the California and Chilean Primary Sites, during Successive Years*

	Maximum observed twig weights ($g\ m^{-2}\ yr^{-1}$)			Litter total ($g\ m^{-2}\ yr^{-1}$)	
	1971	1972	1973	1972	1973
California					
Rhus ovata	614.2	77.6	1,164.8	78.8	380.3
Ceanothus leucodermis	369.4	37.1	335.9	205.5	314.6
Heteromeles arbutifolia	646.1	146.4	708.6	206.2	254.8
Arctostaphylos glauca	1,950.1	673.6	1,935.6	406.5	531.3
Adenostoma fasciculatum	273.5	15.4	463.2	96.6	213.2
Quercus agrifolia[a]	249.6	7.9	461.9	513.1	646.0
Q. dumosa	379.9	25.3	483.9	180.0	117.3
Ceanothus greggii	923.0	87.8	1,057.0	249.7	383.0
Mean	675.7	133.9	826.4	242.1	355.2
Herbs (total)	–	–	11.5		
	1971	1972–73	1973–74	1972–73	1973–74
Chile					
Lithraea caustica	403.2	657.1	627.4	290.3	117.9
Trevoa trinervis	449.2	325.7	443.7	176.2	276.8
Kageneckia oblonga	487.4	727.4	559.9	209.9	121.9
Colliguaya odorifera	410.5	474.0	470.0	427.5	374.8
Satureja gilliesii	212.7	205.2	217.8	155.6	107.1
Cryptocarya alba[a]	794.1	592.5	485.0	326.1	183.4
Quillaja saponaria[a]	276.1	272.4	274.5	132.2	332.2
Mean	433.4	464.9	439.8	245.4	216.3
Herbs (total)	–	105	52		

a. Shrub forms.

as we have seen. Second, in these evergreen species the leaves presumably serve as storage organs from which materials are reabsorbed before they are shed. Leaching of materials from the litter also reduces their biomass. Finally, herbivores take some toll of the primary production.

Nutrient Relations

Although complete data on the nutrient relations of the evergreen-scrub communities of California and Chile are not yet available, it is clear that the two systems differ in their soil-nutrient store of both nitrogen and phosphorus, as has been discussed in Chapter 3. Estimates of the store of nitrogen in the soils in Chile average about 250 g m^{-2}, which is somewhat above one-half of the 440 g m^{-2} estimated for the Californian soils (estimates are for the upper meter of soil). At the same time, the soils of the Californian sites carry less phosphorus than their Chilean counterparts. These differences are not totally reflected in the nutrient content of the leaves of these communities. The average percentage nitrogen of old and new leaves in the two communities is

TABLE 5-15 *Annual Production of Dry Matter, Based on Gas-Exchange Measurements, for Three Dominants of the Evergreen Community at the Californian Primary Site[a]*

Species	Net photosynthesis (g CO dm^{-2} yr^{-1}) (leaf surface)	Net photosynthesis (g CO$_2$ g dw yr^{-1})	LAI (m^2 m^{-2})	Leaf weight (g m^{-2})	Net shrub photosynthesis (g CO$_2$ m^{-2} yr^{-1}) (ground surface)	Shrub production (g organic matter m^{-2} yr^{-1})
California						
Rhus ovata	2,430		1.95		4,738.5	2,909.4
Heteromeles arbutifolia	2,667		2.81		7,494.3	4,601.5
Adenostoma fasciculatum		14.16		290.8	4,117.7	2,528.3

a. Gas-exchange measurements are from Dunn (1970). Factor of 0.614 used for g CO$_2$ to g dry-matter conversion.

TABLE 5-16 *Litter Production of the Dominant Plants of the Evergreen Communities at the Californian and Chilean Primary Sites*

Species	Litter production (g m^2)						Percentage of total				
	Leaves	Stems	Bark	Flowers	Fruits	Total	Leaves	Stems	Bark	Flowers	Fruits
California (1973)											
Rhus ovata	342.1	33.6	0.0	4.6	0.0	380.3	90.0	8.8	0.0	1.2	0.0
Ceanothus leucodermis	303.7	10.8	0.0	0.1	0.0	314.6	96.4	3.3	0.0	0.3	0.0
Heteromeles arbutifolia	190.6	8.4	0.0	15.1	40.7	254.8	74.8	3.3	0.0	5.9	16.0
Arctostaphylos glauca	430.6	1.2	53.7	44.8	1.0	531.3	81.0	0.2	10.2	8.5	0.2
Adenostoma fasciculatum	114.7	17.5	4.5	9.6	66.9	213.2	53.8	8.2	2.1	4.5	31.4
Quercus agrifolia[a]	332.7	22.6	278.5	3.2	9.1	646.0	51.5	3.5	43.1	0.5	1.4
Q. dumosa	96.8	15.6	0.1	2.5	2.4	117.3	82.5	13.3	0.1	2.1	2.0
Ceanothus greggii	371.8	10.8	0.0	1.0	0.0	383.0	97.1	2.6	0.0	0.3	0.0
Mean	272.9	15.1	42.1	10.1	15.0	355.2	78.4	5.4	6.9	2.9	6.4
Chile (1972–73)											
Lithraea caustica	245.8	18.7		5.5	20.3	290.3	84.7	6.4	—	1.9	7.0
Trevoa trinervis	142.3	27.5		4.0	2.4	176.2	80.8	15.6	—	2.3	1.4
Kageneckia oblonga	198.1	9.5		0.8	1.5	209.9	94.4	4.5	—	0.4	0.7
Colliguaya odorifera	365.5	39.9		14.2	7.9	427.5	84.5	9.3	—	3.3	1.8
Satureja gilliesii	132.8	18.7		3.9	0.2	155.6	85.3	12.0	—	2.5	0.1
Cryptocarya alba[a]	305.9	20.1		0.1	0.0	326.1	93.8	6.2	—	0.0	0.0
Quillaja saponaria[a]	102.4	28.1		0.0	1.7	132.2	77.5	21.3	—	0.0	1.2
Mean	213.3	23.2		4.1	4.9	245.4	85.9	10.8	—	1.5	1.7

a. Shrub forms.

135

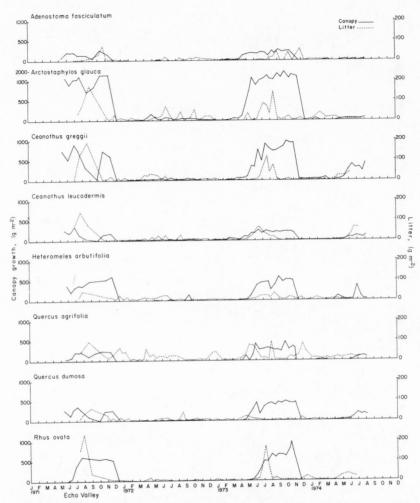

FIGURE 5-21. *Seasonal canopy and litter biomass of the dominant plants of the Californian evergreen community. The values for litter are the amounts collected on the dates indicated. The sampling periods were generally at two-week intervals. The canopy figures are the average dry weight of current twig growth on the plant at the indicated time (Mooney and Kummerow, unpublished data).*

comparable (Table 5-17). The average phosphorus content of the leaves of the Californian plants, however, is only one-half that of the Chilean plants.

The average ratio of nitrogen to phosphorus in leaf material is only 8:1 in the Chilean leaves and up to 15:1 in the leaves of the Californian evergreens. The latter value is approximately that found in the leaves of the Mediterranean evergreen-garrigue community (Lossaint, 1973), but is much lower than the

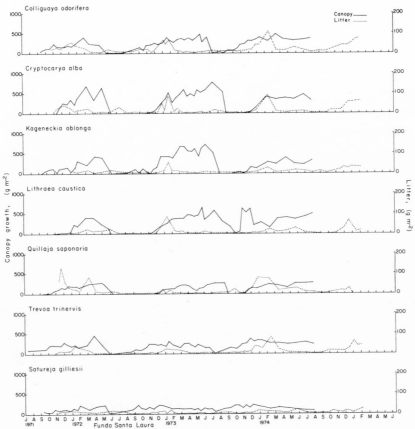

FIGURE 5-22. *Seasonal canopy and litter biomass of the dominant plants of the Chilean evergreen community. The values for litter are the amounts collected on the dates indicated. The sampling periods were generally at two-week intervals. The canopy figures are the average dry weight of current twig growth on the plant at the indicated time (Mooney and Kummerow, unpublished data).*

ratio of 45:1 found in the shrubs of the phosphorus-poor Australian heath community (Specht, 1969).

Since the tissue nitrogen content is similar in the two communities, it can be further estimated that the g m^{-2} of nitrogen in the biomass differs by a factor of 3 (assuming a mean tissue nitrogen content of 1 percent the community values would be 23 g m^{-2} in California and only 7.4 g m^{-2} in Chile). Comparable values for nitrogen content in the garrigue have been measured by Lossaint (1973): 15.6 g m^{-2} in the biomass and 650 g m^{-2} in the soil (Tsutsumi, 1971); and temperate forests show values double the latter amounts (Duvigneaud and Denaeyer-De Smet, 1971).

TABLE 5-17   *Leaf Mineral Content of the Dominant Plants of the Evergreen Communities at the Californian and Chilean Primary Sites*

Species	Percent nitrogen		Percent phosphorus	
	New leaves	Old leaves	New leaves	Old leaves
California (June 1973)				
Rhus ovata	2.54	0.67	0.18	0.06
Ceanothus leucodermis	2.00	1.25	0.15	0.06
Heteromeles arbutifolia	1.28	0.80	0.11	0.04
Arctostaphylos glauca	0.87	0.47	0.08	0.03
Adenostoma fasciculatum	1.16	0.83	0.08	0.05
Quercus agrifolia	1.72	1.17	0.18	0.09
Q. dumosa	1.63	0.96	0.13	0.07
Ceanothus greggii	1.67	1.14	0.09	0.04
Mean value	1.61	0.91	0.13	0.06
Mean N/P ratio	12.4	15.2		
Chile (December 1973)				
Lithraea caustica	–	0.65[a]	–	0.08[a]
Trevoa trinervis	2.92	–	0.11	–
Kageneckia oblonga	2.45	0.98	0.33	0.18
Colliguaya odorifera	1.16	0.98	0.13	0.09
Satureja gilliesii	0.93	–	0.16	–
Cryptocarya alba[b]	1.96	0.73	0.28	0.07
Quillaja saponaria[b]	1.42	0.75	0.22	0.06
Mean value	1.81	0.82	0.21	0.10
Mean N/P ratio	8.6	8.2		

a. April 1973.
b. Shrub forms.

Summary of Biomass Distribution and Net Productivity

The shrubs of the evergreen communities of California and Chile are comparable in a number of parameters. They have similar volumes, ca. 3 m^3, which are greater than those of shrubs from more arid sites. They both have a comparatively large proportion of their aboveground biomass in leaves (20 percent) and, yet, a low leaf-area index (ca. 2). The average features of the leaves of the dominants of these communities are virtually identical, yet they differ from those of evergreens from other climates. The leaves in both cases have a very high average specific weight (19 mg cm^{-2}) and a high sclerophyll index (ca. 260). These results indicate that the shrubs in the two communities apportion their structural carbon in a very similar manner.

The nitrogen contents of the leaves of the shrubs of the two communities are similar, but phosphorus levels are substantially lower in the Californian plants, owing to low stores in the soil.

The aboveground shrub biomass and productivity in Chile are only one-third those of the Californian community, owing chiefly to the lower plant cover in Chile. The ratio of aboveground biomass to twig production is, however, very similar in the two communities (ca. 3:1). Furthermore, the distribution of the components of twig production is quite similar (ca. 70 percent leaves, 20 percent stems, 10 percent reproductive parts).

Productivity in the Chilean community was fairly constant throughout the study, whereas in California there was a sixfold difference between years in response to a variable rainfall.

SUMMARY: PRODUCERS AND THEIR ADAPTIVE RESPONSES

The evidence for evolutionary convergence between the plants of the mediterranean ecosystems of California and Chile is overwhelming despite the taxonomic dissimilarity between the woody plants of the two regions. The modal morphological types found at matched sites on the two continents are similar, but different from those found at adjacent, climatically dissimilar sites in either area. This is true even when one compares climatic extremes, such as mediterranean coastal vs. interior or desert vs. montane—or at the fine level of particular topographic positions within a single climatic type. What are the mechanisms that have led to this phenomenon?

We found that at any given position on the climatic gradient whether in California and Chile, the plant types present have characteristics that optimize carbon gain. Features such as leaf size, leaf-area index, leaf duration, specific leaf weight, and photosynthetic pathway, which relate to photosynthetic adaptiveness, are comparable between the dominant plants occupying similar climatic areas in California and Chile. One example of the relationships between climate, optimal form, and photosynthesis has shown, by a simple cost/benefit (leaf production carbon gain) analysis, that evergreen species are favored in regions where photosynthesis is restricted by drought or temperature for periods less than about three months. Where there are environmental restrictions for greater periods, deciduous elements are favored (Miller and Mooney, 1974). These constraints are realized in both California and Chile: the gradient of drought-deciduous types dominates in areas where drought restricts photosynthesis for long periods; that of winter-deciduous types dominates where cold restricts photosynthesis. Broadleaf evergreens predominate between these limits.

The factors selecting for optimal productive systems in any environment are indicated in Figure 5-23. These factors are also those that regulate carbon gain in the community. For the evergreen shrubs of the mediterranean-climatic gradient, water availability is the predominant selective force (Mooney, Harrison, and Morrow, 1975).

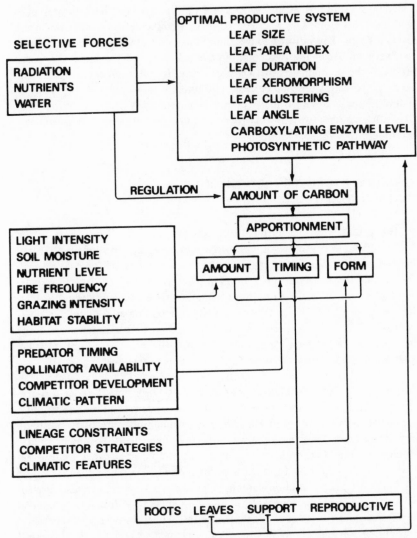

FIGURE 5-23. *Factors selecting for the timing, form, and amount of carbon apportionment in plants (Mooney, 1975).*

If the selective and regulatory constraints on productive systems are the same under equal climates, then productivity should be equal. This was found to be the case, at least potentially, in the scrub communities of both California and Chile. The seasonal production of leafy twigs and litter was virtually identical on a unit-biomass basis. Thus, the environment selects not only for the configuration of the productive system, but also, of course, for its seasonal output. This, then, opens the possibility of convergence in function in other features of the ecosystem.

The evolutionary selection of the timing, amount, and form of the carbon apportioned to the roots, leaves, support, and reproductive systems in any plant is determined by many factors, some of which are indicated in Figure 5-23. Under equal selective forces, equal apportionment patterns should prevail.

As discussed earlier, evergreen shrubs fix carbon the year round, although at reduced rates during the winter and especially during the summer drought. The apportionment of this carbon to various functions, however, is not uniform through time. For example, canopy construction, flowering, and root development all occur at different times. Apparently, carbon limitations preclude allocation to all of these activities simultaneously. This pattern has numerous ramifications in the functioning of the total ecosystem. For example, new energy sources for consumers are pulsed into the system, not made available continuously. Furthermore, most of the plants utilizing a given resource base in a closed-canopy community, e.g., evergreen shrubs, will be synchronous in certain of their activities. Canopy development on all shrubs ensues when tissue damage by freezing is no longer probable. An undue delay in canopy development may result in a competitive loss to overtopping by plants of other species. This synchronous competitive race for the construction of canopies results in an "herbivore window," owing to the rapid construction of leaf tissue that is low in specific weight as well as in content of secondary chemicals.

The selective forces operating to control apportionment timing in these plants are diverse (Figure 5-23), and they vary in intensity during the course of the year. These forces operate in the control of seasonal carbon partitioning to the various structural and chemical components of the plant.

In any event, in comparable climates the same factors operate in selecting and controlling seasonality. In the evergreen-scrub communities of both California and Chile the canopy growth of the herbs proceeds during winter and early spring, that of the shrubs principally during spring. The greater variability of the shrub growth period in Chile may be due in part to the community's more open nature, and thus to a somewhat lower degree of selection for synchronization. The climatic differences between the two areas no doubt also play a role in these growth-pattern differences.

The amount of carbon apportioned to the various plant parts in any region is under the selective control of numerous factors (Figure 5-23). In the mediterranean regions of California and Chile, soil moisture and fire frequency have probably been the principal factors. Both select for apportionment to root over shoots. In both California and Chile there are deeply rooted shrubs, most of which fire-sprout. These features are thus convergent in response to comparable selective factors.

Finally, the form the carbon skeleton assumes in a plant is the result of complex selection. The climate selects directly for form, by determining the optimal productive system, as discussed above. Thus, comparable climates, if each supports a closed community, will support dominants with similar canopy types. Other morphological features, such as the productive system, will also be convergent, since they may respond directly to climatic selection as well.

Thus, in answer to the question posed earlier, "How does convergence operate?", it appears that the selective forces in California and Chile are comparable at the producer level. This results in the evolution of equivalent forms in the two regions, forms that in optimizing carbon gain for this climatic type take on comparable potential productivities. Because of both internal and external constraints, the carbon comes to be allocated in a similar manner at specific times to specific functions within a given growth-form type. These patterns of carbon allocation in turn dictate a given temporal and spatial partitioning of energy and biomass within communities evolving under equal climates.

The forces leading to the evolution of optimal forms are indeed strong, as is indicated by the sorting out of equivalent types in California and Chile along topographic gradients encompassing small geographic distances. The fact that greater similarities in form can be found between plants occupying matched habitats in California and Chile than between more closely related species occupying the general mediterranean climatic region in either area further indicates the strength of natural selection (Parsons, 1973).

There are, however, a number of differences between the vegetation forms of the Californian and Chilean primary sites that do not support the convergence hypothesis, and thus warrant close attention. First, the vegetation of Chile is much more open and more diverse than its Californian counterpart. Very likely, the explanation for these differences lies not in evolutionary factors but in differences in recent land-use patterns. Although the study sites in California and Chile were matched as closely as possible in terms of climate, landforms, and land use, differences do exist, particularly in land use. The Californian site, as is typical of most of the vegetation in the region, is perturbed at frequent intervals by devastating fires, which in recent times have probably increased in intensity while decreasing in frequency. At the same time, man has had little *direct* impact on this vegetation, now or in the past. In Chile, however, man has had a massive direct impact on the vegetation, with results perhaps not as dramatic as his ravages in North Africa (Mikesell, 1960), but yet quite devastating. Large areas of the mediterranean-climatic regions of Chile today are hillsides supporting a very sparse vegetation composed almost entirely of spiny or toxic plants. There is little doubt that these regions once supported a more fully developed vegetation.

The Chilean primary site represents one of the least disturbed vegetations in central Chile, yet, as we have shown, it has had a long history of man-induced perturbations, most of which were highly localized within the site—e.g., selective woodchopping, charcoal-making, plot-clearing, grazing, mining. These disturbances have resulted in a complex mosaic of small land units having various histories and exhibiting various successional states. Recovery rates in these arid climates are slow, and site degradation has been continuous for many decades and even centuries. This microsite complexity, then, might well explain the lesser degree of cover and the greater within-site α-diversity in Chile. Gulmon (1976) provides additional evidence for this viewpoint. She found that mediterranean weedy annuals form more diverse annual associations in Chile than in California at points of equal climate. She indicates that site degradation due to the prevailing grazing practices in Chile has led to the greater herb diversity there.

The opening of the vegetation by a series of point disturbances, rather than periodic base leveling by fire, has other impacts on the vegetation structure in Chile which makes it differ from its Californian counterpart. An herbaceous component has developed within the woody vegetation of Chile, a great portion of it consisting of weedy annuals. At the same time, no post-burn herbaceous native flora exists at present in Chile, in contrast to the rich post-burn herbaceous flora of California. Either it never existed in Chile—perhaps because frequent, large-scale, wind-driven fires, and their concomitant evolutionary stimulus, have apparently not existed—or such a flora was simply swamped out by the weedy invaders.

There is a second, perhaps more important difference between the vegetations of California and Chile that cannot be explained simply by dissimilarities in land use. The dominants of the Californian vegetation undertake a synchronized growth period in the spring. In Chile, by contrast, the growth of the dominants as well as the timing of other phenological events, although centered in the spring, is more protracted. This disparity could be due to the somewhat greater climatic equitability of the Chilean primary site. Further, the greater openness of the Chilean vegetation would result in less selection for synchronous canopy growth.

It would appear, in any case, that the components of the vegetation in California and Chile are individually quite convergent. The particular configurations they assume today, in forming the two communities, are rather different, chiefly because of dissimilarities in land-use history. It may be asked whether, prior to the Conquest, even the gross configurations were convergent.

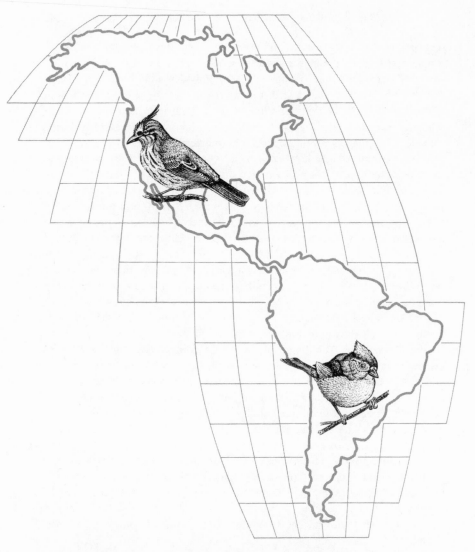

Convergent Evolution in the Consumer Organisms of Mediterranean Chile and California

M. L. Cody
E. R. Fuentes
W. Glanz
J. H. Hunt
A. R. Moldenke

Earlier chapters in this volume have described the physical setting (climate, land forms, and soils) and the resultant vegetation types (the morphology, phenology, and physiognomy of the plants) of the mediterranean zones of Chile and California. The likely origins of the biota, and the possible evolutionary histories of the better known organisms, were also discussed at an earlier stage. In this chapter we shall summarize the evolutionary responses of the consumer organisms to these physical and vegetational factors, and discuss the extent to which intercontinental similarities in such factors have produced parallel or convergently similar adaptations in the consumers.

BASIC CONSIDERATIONS IN CONVERGENT EVOLUTION

Our basic thesis is that similar adaptive landscapes (in the sense of Sewall Wright) will produce similar adaptive responses in unrelated organisms. Such adaptation should be demonstrable not only at the level of the organism, but also at the suborganism and supraorganism levels. Although we shall not be much concerned here with suborganism evolution, investigation of the way in which natural selection arranges complementary species into stable, invasion-resistant sets called *communities* is a major goal of our work. Thus, in the sort of intercontinental study we undertook, there are several considerations of general importance we must evaluate: (1) Are the adaptive landscapes really similar? (2) Are the populations confronted by these landscapes sufficiently different in systematic or taxonomic origins that adaptive similarities can be attributed to convergent evolution rather than to common ancestry? (3) Given similar adaptive landscapes and substantially different gene pools on two continents, is convergent evolution a reality? That is, are similar, quasi-optimal solutions to adaptive problems in different mediterranean zones reached consistently or only infrequently, and are such solutions obscured by the existence of a variety of alternative optimal solutions? Finally, (4) are the remaining inevitable differences between analogous organisms, populations, and the communities they constitute themselves explicable in terms of evolutionary or ecological theory, with reference perhaps to historical, temporal, or other differences in the selective regimes of the two continents?

The material necessary for a reasonably confident affirmative reply to (1) precedes this chapter. The initial choice of the mediterranean zones of Chile and California, based as it was on rather preliminary impressions, has been confirmed by subsequent data collection as being particularly appropriate for a study of the depth and magnitude we sought and attained. Clearly, with increasingly fine scales of measurement for both physical and biological variables, more and more differences between the continents will be unearthed. But for the purposes of predicting and demonstrating convergent evolution, we can rely on extensive similarities in the adaptive landscapes of

Chile and California; we can, as well, quantify and assess any differences that might exist between the two regions.

Question (2) was discussed in Chapter 2 of this volume, where evidence was summarized to affirm the long-separate and distinct evolutionary histories of North and South American vertebrates. Although a similar conclusion about the evolutionary histories of the invertebrate fauna of the two continents cannot be evidenced, it is presumed warranted by their wide historical separation. The origins and evolutionary histories of the faunas of Chile and California are reflected in the modern systematics of these faunas, and question (2) can be answered by a quick check of the taxonomic affiliations of the organisms we found on our study sites. Most of the animals in the three vertebrate groups studied—lizards, birds, and mammals—show substantial differences at the level of the family, i.e., the dominant families on one continent are absent or poorly represented on the other continent. Moreover, there are few genera or species in common and, as with the plants, the only taxa found in both places are cosmopolitan "weeds" such as rats, vultures, and honeybees. Of the insects studied—mainly ants and plant pollinators (including in particular bees)—rather closer taxonomic parallels were noted. This might be attributable to large-scale, passive dispersal in insects, but seems more likely an artifact of insect vs. vertebrate taxonomy—the vertebrates have been rather more finely divided, by more empathetic taxonomists. The ants and bees in the two regions share a good many of the same genera but just one or two species. The pollinator insects in general show a similar intercontinental familial representation, but with prominent exceptions: the hoverflies (Syrphidae), for example, are numerically far more important in Chile; and the species-diverse Andreninae of California are replaced in Chile by the genus-diverse Colletidae. In sum, any attempt to assign for any pair of consumer groups a one-to-one species analogy or correspondence will certainly involve different species, almost always different genera, and more likely than not representatives of different subfamilies or families. Community-level analogs, of course, will routinely include the wide range of differential familial representation already mentioned.

Though questions (1) and (2) will be affirmed or qualified where appropriate in this chapter, we shall be concerned chiefly with questions (3) and (4). The extent to which unrelated species and species assemblages have come to resemble each other on the two continents can be assessed with the data in hand, and can be tested independently with different taxa and different consumer trophic levels. Theoretically, the number of such tests we can perform is large, for not only do we consider several broadly or narrowly different taxa—in groups of ecologically related species within which competition is expected as a dominant, order-creating force—but we can draw upon several pairs of matched study sites on the two continents, and our comparisons sometimes extend over different sets of years with different local climatic conditions.

The organization of our results on consumer organisms follows from basic considerations of community structure. Any community can be viewed, following the pivotal paper of R. H. MacArthur (1970; see also R. H. Mac-Arthur *Geographical Ecology,* 1972), as a group of "utilization curves," a set of related resources for which the species of the community compete either potentially or actually. This representation (see Figure, 6-1) implies a great deal of what is important about the community, including niche relations, overlaps and breadths, information on invasibility and diversity, and implications for the stability of the community.

Most of the more solid inferences we can draw on the convergent evolution of solutions to adaptive problems will concern the adaptive nature of species packing and resource use in communities such as that epitomized in Figure 6-1, and will be affected by assumptions on the possible equilibrium

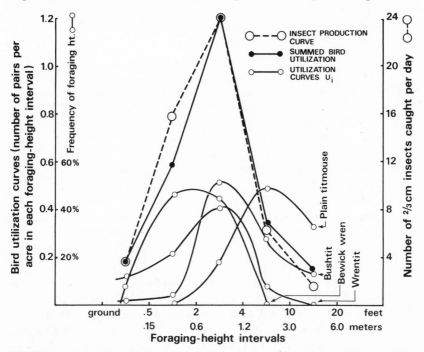

FIGURE 6-1.   *Utilization curves of four species of insectivorous birds of Californian chaparral, given as frequency distributions (left-hand ordinate) over feeding-height intervals (abscissa). The sum of the utilization curves of this community, shown in pairs per acre (0.4/ha) for each height interval, is obtained by multiplying each species' utilization curve by its density in pairs per acre, and summing over species. By adjusting the right-hand ordinate scale, we can match the summed and weighted utilization curves to the insect-production curve. P, estimated by the number of 0.7-cm insects caught on Tanglefoot plaques exposed in each height interval for 24 hours.*

of such species assemblages. Equilibrium of this sort is in turn affected by the availability of various types of species (as represented by their utilization curves), the densities of these species, and the order in which they have arrived, over time, to exploit the resource gradient. A community such as that of Figure 6-1 is at equilibrium if the sum of the several species' utilization curves weighted by species densities is the best least-square fit to the productive curve for the resource gradient. The best fit, and therefore the most internally consistent community, is thus described in terms of (a) a set of species utilization curves and (b) a set of species densities; together these two data sets, or vectors, define the optimal combinations of species for the community, in terms of the species that should occur and the abundances they should occur in.

Production over the resource gradient (e.g., grams of food of particular sizes) is some function P, which in its simplest form represents a plot of the level of each resource type. Thus, if X_i is a row vector of species densities and U_i a column vector of utilization curves, the X_i will be adjusted through competition and natural selection in such a way that $(P - \Sigma_i X_i U_i)^2$ is minimal, and constitutes the smallest possible amount of unused resource.

For convergent evolution to have occurred, there should be correspondence *at the species level* in the shape and positioning of utilization curves (given that the resource gradients at each of two compared sites correspond), and *at the community level* in the number, relative juxtapositioning, and relative abundances of species (given that the production curves for the resource gradients likewise correspond). For the purposes of this chapter, we select the type of community represented by Figure 6-1 as a baseline unit for intercontinental comparisons.

The heart of the chapter comprises two major sections, "Patterns of Diversity and Distribution" and "Patterns of Density and Resource Use." The first of these investigates the similarity of species-packing patterns within communities and at levels beyond the community. That is, we look at various levels of species diversity, from diversity in communities found within small patches of homogeneous habitat to diversity between different habitat types and between different geographic regions within each of the continents. Different aspects of species diversity have been given different names: (1) the diversity of species within a particular habitat patch, from 2-5/ha for birds to perhaps 200-500/m^2 for ants, is called α-diversity; (2) the rate of species change, or turnover, per unit change of habitat, is called β-diversity; and (3) the additional diversity accrued where species change intracontinentally with shifts over geographic area, uncorrelated with habitat changes or nearly so, is called γ-diversity.

It will be seen that α-diversity is quite predictable from the structure of the habitat for many groups of organisms, and is therefore predictable from competition and species-packing theory. Because similar competitive processes operate in all natural communities, striking similarities in species-packing levels, the α-diversities, are observed between structurally similar habitats on different continents. But the turnover rates of species between habitat types, the β-diversities, depend far less on competition, far more on historical, topographical and chance considerations. The latter factors affect, in taxon-specific ways, the availability of different numbers of species, the speciation rates, the relative accessibility of different habitat types, and the opportunities for dispersal through and adaptive radiation within habitat constellations. The β-diversities are thus far less predictable from ecological theory. Both β-and γ-diversity for a given taxonomic group can be crudely correlated to topographic and other geographic aspects of the countryside surrounding the study sites (Cody, 1975); but since these physical conditions inevitably differ between continents, considerable intercontinental differences in both β- and γ-diversity may exist.

The second major section of the chapter looks at patterns of resource use from the community level down through species subsets, or *guilds*, in which competition is expected to be a particularly prevalent factor, to one-to-one species correspondence and the shape and position of individual utilization curves. Here we are interested not only in the number of species that use a particular set of resources, but in the density of each species and in the density of the species set as a whole. We shall comment on the process of *density compensation* (MacArthur et al., 1973), by which the rarity or absence of one or more species in a guild of ecologically related, competing consumers can be compensated by a higher density in one or more of the remaining species. Thus, even though species numbers per guild might not exactly match between continents on parallel resource gradients, perhaps for chance or historical reasons, we nonetheless expect and predict that the densities of corresponding multispecies guilds on the two continents will be more nearly parallel than the average densities in the one-to-one species analogs that comprise these guilds (Cody, 1974).

Density compensation takes place only to the degree that the resource-utilization curves of different species overlap, either potentially or actually. At times, very distantly related consumers, such as vertebrates and invertebrates, share the same resource (e.g., ants and birds often eat seeds of comparable sizes); thus, density compensation can occur within a set of consumers that includes very different organisms (low ant abundance might permit greater seed-eating bird abundance or greater rodent densities, and vice versa). The extent to which such intertaxon compensation takes place will be discussed at the close of section on density and resource use.

While in theory we aim to answer just the same sorts of questions with both vertebrates and invertebrates, inevitably differences of approach, resolution, and emphasis arise that are attributable to a variety of causes: our knowledge of systematics and of natural history is uneven within the invertebrates and between invertebrates and vertebrates, and the number of species involved is several orders of magnitude greater in inverbebrates than in vertebrate animals. We simply cannot collect similar data and conduct similar tests on 2,000 bee species as on a dozen lizard species. While we can make niche comparisons in several resource dimensions in lizards and accurately align niches between analogous species and assess density compensation, with pollinating insects we must constrain our conclusions to generalities about relative diversity and the broader patterns of resource division; much more detailed information at the species level would be required in the latter group before a more comparable and sophisticated analysis can be done.

Although the study we undertook was a large-scale enterprise involving many people, we were far from exhaustive in studies on consumer organisms. Certain taxonomically and, we hope, ecologically discrete groups were selected for study, but others were ignored. Our data are quite complete for the consumers of flowering parts, seeds, and insects—i.e., ants, pollinator insects, lizards, birds, and mammals. We know much less about other potentially important and extremely diverse consumer groups, such as herbivorous insects and litter and soil organisms. Top predators drew scant attention, since their much lower densities make data collection more difficult and ecological studies less rewarding. Though we compiled approximate censuses of the bird and mammal carnivores, and know something of the overall biomass of insect predators, we shall draw no precise cross-continental comparisons.

At the close of the chapter such generalizations as our studies permit will be drawn and set in perspective. There we shall emphasize not only that convergent evolution can be demonstrated at various levels of species organization and provide us perhaps the most vivid evidence for the reality and strengh of natural selection, but also that conclusions of equal evolutionary and ecological importance can be drawn from the apparent exceptions to the convergence trend. It should be borne in mind that many of our conclusions will be greatly strengthened when further studies have included third and fourth continents as points of comparison. Attempts to draw firm conclusions from comparisons between just two of the world's five mediterranean zones suffer much the same difficulty as the draftsman seeking to fit a definitive curve through just two points in a plane. Some of us, indeed, have broadened our perspectives by conducting parallel studies in the Old World mediterranean zones of Sardinia and the Cape region of South Africa.

This volume is intended as a companion to the volume of facts and figures (Thrower and Bradbury, 1977) that serves as its data base. Accordingly, the synthesis and conclusions presented here offer few of our original data, but refer the reader instead to the data atlas. Similarly, we omit a good deal of the intermediate analysis, referring the reader to published articles for such details.

Finally, it should be stressed that this chapter is the work of all of the scientists engaged in consumer studies in the Convergent Ecosystems integrated-research project. Some of these studies have yet to be completed, and others have yet to be fully analyzed. Thus, although the chapter presents a synthesis of the results so far available, it does not by any means represent a final work, and many further contributions are certain to follow.

PATTERNS OF DIVERSITY AND DISTRIBUTION

In the first two parts of this section we discuss the various sorts of diversity and distributional relations in consumer organisms whereby the corresponding Californian and Chilean populations can be compared, and in later sections we extend the comparisons to cover individually the various groups of consumers studied.

Levels of Diversity and Species-Area Curves

We begin with the standard species-area curve of the biogeographer, which plots numbers of species S in a limited taxonomic group against sample area A. If both axes are logarithmic, such plots are usually straight lines $S = cA^z$ or $\log S = \log c + z \log A$. For nested samples of mainlands, the values of z are around 0.1–0.2; c is a constant. For Chilean and Californian birds (Figure 6-2), using sample areas centered in the chaparral of the mediterranean zones of the two countries, the z exponents, or slopes, are both 0.13—not detectably different between continents.

Such graphs, when plotted in detail, describe not only this broad relation between species number and area (which must be accorded, at this time, only tentative significance), but also specific diversity relations of more definite ecological significance. Since bird-study areas that include all of the "appropriate" species and exclude structurally different vegetation types are determined empirically at ±2 ha (see arrow on graph, Figure 6-2), the number of species at that level of resolution represents species-packing level, or α-diversity, for the chaparral and matorral of primary community sites. This sample area lies just where the two curves of Figure 6-2 are most similar, at around 10^{-2} mi^2 (1 mi^2 = 2.6 km^2). When increasingly smaller areas are sampled in the two chaparral primary sites, diversity decreases more rapidly in Chile than in California, and this extrapolation is practical down to areas around 15 × 15 m^2, in which species number, or, here, *point diversity,* is significantly greater in California than in Chile. Thus, there is less interspecific territorial overlap in the Chilean birds, from which we might predict (assuming complementarity between habitat use and food use, following Cody, 1974) that the Chilean birds should be more food-generalist, as proves to be the case (see below, p. 176).

Proceeding to the right from the arrow that marks the chaparral primary sites (in Figure 6-2), we see that increases in sample area effectively increase habitat diversity. And since the topography and therefore the habitat diversity of the Chilean primary site are similar to those of the Californian primary site, the slopes of the curves to the right of the arrow give the rate of accumulation of bird species with accumulation of habitat types. These slopes, then, measure β-diversity, and we can see from Figure 6-2 that β-diversity is greater

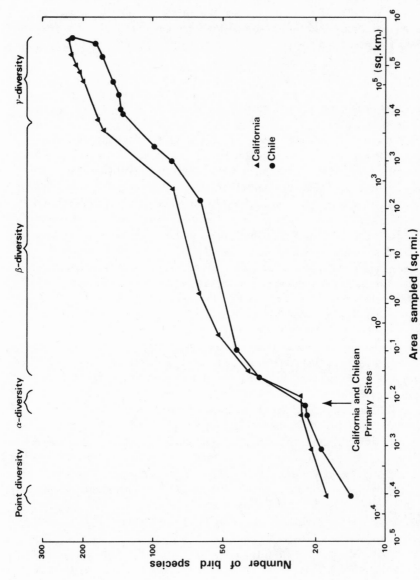

FIGURE 6-2. *Species-area curves for Californian and Chilean mediterranean birds, centered in the primary-site chaparral. Various sections of the curves indicate various diversity measures.*

in California than in Chile, since by around 10^4 mi^2, when most of the additional habitat types have been accumulated, 30 percent more species, 177 vs. 136, have been found in California.

Increases in sample area beyond a hundred miles (160 km) of the chaparral site (area 10^4 mi^2) do not encompass much new habitat, but a few new bird species are found, nonetheless. Accumulation of this sort constitutes chiefly geographic replacement, in similar habitats, by many pairs or larger groups of species. Over this stretch of the abscissa, Chile regains almost all of the species by which it fell short of the California total in the central, β-diversity, portion of the graph. The resulting bird-species totals, 235 vs. 230, are remarkably close, owing to the inverse but opposite relations between β- and γ-diversity in the two countries: California, via-à-vis Chile, is higher in β-diversity, lower in γ-diversity.

Treated in this way, species-area curves furnish a considerable amount of information about species packing within habitats and about species turnovers between habitats and geographic locations; after a fashion, the curves also summarize the diversity and distributional data that characterize a particular region. For the birds of the two mediterranean zones, the relative diversity values are shown in Figure 6–3. Equality in species number is observed for α-diversity and for total species counts: slopes less than unity indicate higher diversity in Chile (γ-diversity); slopes greater than unity indicate higher counts in California (point diversity and β-diversity).

Diversity, Density, and Niche Characteristics

As with the diversity and species-area curves, there are some simple and obvious relations to be expected between diversity and niche characteristics. We may define the niche of a species as the combination of its utilization curves over however many resource dimensions are important in the interactions (actual or partial) between the species and its competitors. In practice, the important niche resource dimensions are quite few, usually including only food type, foraging site, habitat type, and perhaps time.

A niche can be broad or narrow in each resource dimension, indicating that the species is either generalized or specialized, respectively, in the use of that resource. Another variable is the extent to which the niches of adjacent species may overlap; but niche overlap is most likely determined by environmental predictability (Cody, 1974, cf. May and MacArthur, 1972), and therefore is expected to be constant between mediterranean sites that have been selected in advance for their climatic similarity.

Niche breadth and species densities interact in various ways to yield diversity correlations. Point diversity, for example, precludes consideration of habitat as a variable resource. We therefore expect that higher point diversity will be achieved by species with narrower niches over food type or foraging site, and we expect to find species that are food or foraging-site generalists

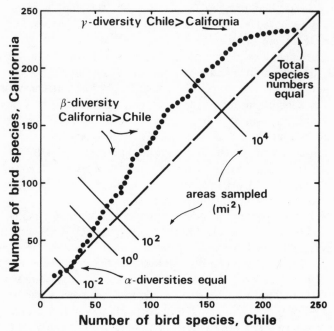

FIGURE 6–3. *Relative species numbers and species diversities in California and Chile, for birds centered in mediterranean sclerophyll scrub. The values are equal between continents at two points: α-diversity and species total. Elsewhere the slope of the relation is greater than 1 (β-diversity greater in California) or less than 1 (γ-diversity greater in Chile). Values given are the area sampled in square miles (1 mi² = 2.6 km²).*

at points of low diversity (given comparable food and feeding-site ranges, as in matched points between Chile and California).

Since species can subdivide habitat within study sites of a few acres, α-diversity is less clearly correlated to changing niche breadths. But as before, fewer species, each generalized in both food and habitat use, can be supported locally and will result in reduced α-diversity, even though each such species can be common. Specialized species, by contrast, having narrow, low-variance utilization curves, are not expected to be as common as generalized species, especially if production curves are reasonably level across the resource gradient. This relationship follows because the X_i must be smaller in specialized species in order to fit the same production curve P. Thus, specialized species in general would produce higher α-diversities, and each such species would be relatively scarce. Equal α-diversities could be exhibited at two matched sites in which species at one site were generalized in one dimension, such as food type, and specialized in another dimension, such as habitat, while the converse obtained at the other site. Because of the expected (and observed) similarity in permissible niche overlaps for species at climatically similar mediterranean sites, differences in α-diversities are likely to involve differences in niche breadths, with the lower niche breadths occurring in the more diverse sites.

A sequence of habitat types, such as that from grassland through sclerophyll scrub to evergreen woodland associated with increasing moisture availability in mediterranean zones (Mooney et al., 1970; Cody, 1975; Chapter 5), forms a particular sort of resource gradient (see below). The distribution of species over such a gradient yields habitat niche breadth and determines β-diversity. Notice that, although environmental change may occur at different rates for different taxa (e.g., ants and birds) along a particular habitat gradient, such scale or resolution differences between taxa will impair only diversity comparisons between taxa; the comparisons of β-diversity between continents for any given taxon are not affected. Indeed, β-diversity can be low if niche breadths over habitats are large; and for a given limiting α-diversity, higher β-diversity means a reduced habitat niche breadth. It is tautologically true that habitat generalists will produce low β-diversities and that food specialists will produce high α-diversities; it is not so clear that habitat generalists will reduce α-diversities or that food specialists will increase β-diversities, though both suggestions seem intuitively reasonable.

At the highest levels of diversity, γ-diversity and total species count, we are dealing with measures that appear to be largely outside the effects of competition for limiting resources, and into the realms of history, geography per se, chance, and the largely undetermined but certainly complex factors affecting the production, dispersal, and coexistence of species over large geographic areas. We might guess that although the competitive process is similar between continents and among diverse groups of different consumers, the multitude of factors in species production and dispersion must differ drastically both between continents and between taxa; we can expect, therefore, little in the way of generalizations at these higher diversity levels. But the comparisons we undertake must illuminate the similarities and differences in such diversity measures, and must inevitably increase our understanding of them.

Ants

The number of ant species was assessed at each of eight sites, four on each continent. The number of species was lowest in coastal succulent scrub, increased through coastal scrub to peak in the tall evergreen sclerophyll scrub (chaparral) of the primary sites, and decreased again in the montane forest. This pattern describes both the Chilean and Californian transects, and probably reflects the fact that the tall scrub embraces both the open habitat patches of the coastal sites and the sheltered, wooded patches of the montane forest, and hence encompasses the greatest diversity of ant microhabitats.

The number of species per site is uniformly greater in California (nine, fifteen, fourty-five, and twenty-one species, respectively), from 1-2/3 to 3 times the numbers found in Chile (six, nine, fifteen, and ten, respectively). These numbers overestimate ant α-diversity, since species are strongly micro-habitat-specific within each site and therefore include a β-diversity component. The primary sites support three main microhabitat types: evergreen woodland, sclerophyll scrub, and open, disturbed places. In California, twenty-three common species were distributed such that these three subhabi-tats supported eight, twelve, and fourteen species, respectively. In Chile, fourteeen common species occupied the same three microhabitats, the three supporting five, nine, and six species, respectively. Thus, α-diversities are in fact greater in California, by factors between 1-1/4 and 2. Hunt (1973) offers additional details.

Between the two continents, greater similarities than those indicated by these figures do, however, exist. In California there are five species restricted to woodland and five to sclerophyll scrub habitats; in Chile four species are restricted to each of these subhabitats. In addition, at both the Chilean and Californian primary sites, five species occupied both the chaparral and the open subhabitats. The main discrepancy in α-diversities lies in the far greater species number in Californian open habitats (six, vs. one restricted species), a discrepancy to which we shall return later.

Because the Californian and Chilean sites differ more markedly in total numbers of species than in α-diversities, the differences in β-diversities must be even greater. Each of the four sites can be ordinated on a habitat gradient H, such that the structural distance between each two adjacent habitats is given as a certain interval on a habitat axis (see Figure 6–4). This permits β-diversities to be quantified in simple fashion. Species are ranked in a manner reflecting their relative order of entering and leaving the habitat gradient H, such that species early in the ranking are encountered on the habitat gradient at the lowest H values and also leave the gradient earlier, and so on. A census taken at a certain point H_i on the habitat gradient will include species between rank numbers X_i (those found only in habitats with H values higher than H_i) and Y_i (those already lost from the gradient and restricted to habitats with H values lower than H_i). Species turnover between habitats H_i and H_j is then given as $[(X_j - X_i) + (Y_j - Y_i)]/2$, and β-diversity, which is species turnover per unit habitat shift, is given as $[(X_j - X_i) + (Y_j - Y_i)]/2(H_j - H_i)$. In Cali-fornia, β-diversities between each two adjacent study sites, from succulent scrub to montane forest, are twenty, thirty-two, and sixteen species per unit habitat shift, respectively. In Chile, the same transitions are twelve, eight, and nine species per habitat shift, respectively. Thus, β-diversity for mediterranean zone ants, to compare overall figures where M_i and M_j are the beginning and

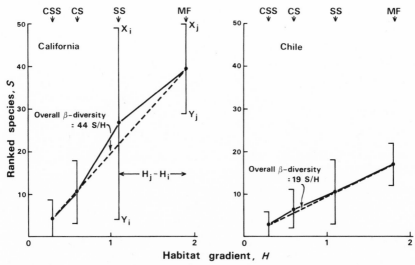

FIGURE 6-4. *Californian and Chilean ant species S ranked according to the order in which they enter and leave (are distributed over) habitat gradient H, censused at four sites each (CSS = coastal succulent scrub; CS = coastal scrub; SS = sclerophyll scrub; MF - montane forest). The value 44 obtained for the California species is over twice the value of 19 obtained from the Chilean censuses.*

end points, respectively, of the habitat gradients, is between 1.7 and 4 times greater in California than in Chile (Figure 6-4; average 2.7 times greater).

The total number of ant species in California is 206 (R. Snelling, personal communication); in Chile the total can be placed at around sixty-five species (Snelling and Hunt, 1975). Since we know the α-diversities of ants for habitat samples around 10^{-2} mi^2 (2.6×10^{-2} km^2) in sclerophyll scrub, the slopes of the species-area curves for increasingly larger area samples, and also the endpoints for the species-area curves (total numbers of ant species), we can calculate the slopes of the curves for the final sections, which represent γ-diversity (see also Figure 6-3). If we estimate the β-diversity slope to be 0.065 in Chile and three times that value in California, the γ-diversity slope in Chile must be three times that in California if the known species totals for the two regions are to be reached. Figure 6-5 illustrates this extrapolation.

For the ants, then, there is a considerable difference between the two continents, a difference that is not excessive at the α-diversity level and may perhaps be even less at the point diversity level, but which becomes greater and greater at higher diversity levels. Within and between habitats there are more ants in California, whereas there is greater ant turnover between geographic localities in Chile. These differences exist despite the fact that similar ant taxa are represented on both continents by species distributed in similar proportions among the taxa.

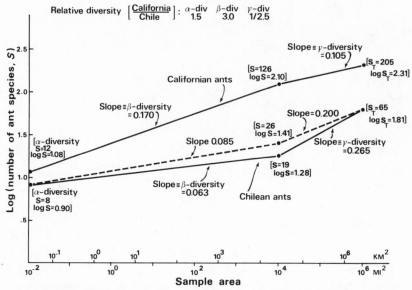

FIGURE 6-5. *Species-area curves for Californian and Chilean ants, beginning in the scrub, where a-diversity in California is an average 1.5 times greater than the corresponding Chilean value. Using the slope of the curve between area = 10^2 and 10^4 mi^2 as an index of β-diversity, and using fixed end points S = 205 in California and S = 65 in Chile, a Californian β-diversity 3 times that of Chile means a γ-diversity in Chile 2.5 times that in California (solid lines). If Chile's β-diversity is assumed to be as much as 0.5 that of California, its γ-diversity is now 2 times that of California (dashed line).*

Pollinator Insects

Although there are diverse forms of insect and other arthropod life that visit flowers and serve as pollinating agents, in this section we shall draw parallels and distinctions strictly between the bees of comparable Californian and Chilean sites. This restriction is both necessary and tolerable; the bee faunas of the two regions are relatively well known, and the bees comprise the single most important group of pollen vectors in the mediterranean regions.

The arid southwestern United States supports ± 2,500 species of bees, arid and semiarid Chile about 800 species, including new species with manuscript names and taxa in the desert and alpine regions awaiting discovery (Moldenke and Neff, 1974; Moldenke and Toro, in prep.; Simpson et al., in press; Wagenknecht, 1966). The remarkable diversity of the Californian and Chilean bee faunas becomes more apparent when these species numbers are compared with the 65 bee speices of southern Florida (Graenicher, 1930) or the 353 bee species of Panama and adjacent middle America (Michener,

1954), two regions noted for extremely high plant diversities, though not large in total area.

Pollinator studies were conducted at five Californian and four Chilean sites: evergreen sclerophyll scrub of the primary sites; two additional pairs of matched sites, in desert scrub and coastal scrub; an early successional post-burn site at Japatul Valley, California, for which there is no analog in Chile; montane forest at Mount Laguna in California, for which the Cerro Roble site in Chile with no understory pollinator-dependent vegetation (see also differences in plant structure and height diversity, Figure 5-4) is an unacceptable analog; and coastal succulent scrub in Chile. For details on these and similar studies the reader is referred to Moldenke (1971, 1975) and Moldenke and Neff (1974).

The pollinators (or, more appropriately, floral herbivores) collected at these sites represented 648 insect taxa (365 bee species) in California, 437 insect taxa (150 bee species) in Chile. The area sampled at each site was normally around 1 km^2, but all observations were confined within the single habitat type that the site typified. The total species count at each site can therefore be termed α-diversity. The values of α-diversity for bees ranged from 25–175 species (100–300 pollinator species in all), and are consequently much higher than the values for other consumer groups. The number of species is lower in the coastal scrub (California 80, Chile 64) and higher in sclerophyll scrub (California 171, Chile 116). These constitute the two best-matched pairs of sites for the bees, and indicate that California is 1-1/4 to 1-1/2 times richer than Chile. But on both continents there is a consistent trend for total numbers of species, total numbers of individuals, and total bee biomass, all increasing from coastal succulent scrub, coastal scrub, desert scrub, and montane forest to a peak in the evergreen sclerophyll scrub. In other California sites, desert scrub at the inland Ocotillo site is intermediate (87 species), successional sclerophyll scrub is significantly poorer than mature sclerophyll scrub (151 species), and the montane forest supports 135 species. In other Chilean sites, desert scrub supports just 29 species, almost certainly because of the infuence of fogs at the coastal El Tofo site, but succulent coastal desert was even poorer (22 species).

In bees, species diversity is clearly not dependent on a simple structural measure, as it is, for example, in birds, but is influenced by climate (e.g., the fogs at El Tofo) or by the understory flowering vegetation (e.g., the presence of only three bee species in the Cerro Roble montane forest understory), and is enhanced by the degree of bare ground available for nest sites and by the warm to hot temperatures propitious for poikilotherm activity. Secondarily, pollinator diversity rises with increasing biomass of the floral resources (scrub > grassland in successional post-burn site > forest understory). Thus, cool maritime climates and continuous canopy forests both reduce sunlight reaching the ground and both therefore inhibit the development of rich bee faunas.

Since α-diversities in California are in excess of those at matched Chilean sites by 25–50 percent and yet local faunal lists are longer by as much as 150 percent, β-diversity must be considerably higher in California. It is rather difficult to disentangle β from γ affects in bees, and the effects of scale are still obscure. It appears that species turnover with habitat is comparable in Chile and California up to perhaps 50 km intersite distance, beyond which Californian species tallies increase at a greater rate than Chilean tallies. When mediterranean Chile is compared to the entire country, species numbers increase from 350 to 800. The same extension from the Californian mediterranean region to include the entire region west of the cordillera from southern Alaska to the southern tip of Baja California (including the Sonoran Desert region of California) increases species number from 1,400 to 1,900 (Moldenke, 1976). Here, species increase in Chile by a factor of 2.3, compared to a factor of 1.4 in Pacific North America; thus, γ-diversity appears be greater in Chile. But since the range of climatic extremes is greater in Chile than in the state of California, the Chilean figure is relatively more compounded by a β-diversity element than is the Californian figure, and the true excess of the Chilean bee γ-diversity can only be guessed. Further studies by Moldenke (1976) and Moldenke, Neff and Toro (unpubl.) elucidate these relations in greater depth.

Lizards

Our lizard studies centered on three matched pairs of habitats on the two continents: coastal scrub, chaparral, and montane woodland (Fuentes, 1974, 1976). On the three Californian sites eight lizard species were found, in six genera and four different families (Iguanidae, Teidae, Anguidae, and Scincidae). Of the nine Chilean species encountered, seven belong to the genus *Liolaemus* in the family Iguanidae. The three habitats are arranged as before from low, dry, and open vegetation in the coastal scrub to tall and mesic forest at higher elevations farther inland. The three habitats in California supported three, five, and five species respectively, compared to three, five, and four in the Chilean sites. Thus, α-diversity between paired sites are virtually identical.

Lizard species, however, select various subhabitats at each site; thus, at the primary sites, ravine evergreen woodland and hillside evergreen scrub were distinguished. In California, four of the five species present were restricted to scrub, and the fifth occurred in both scrub and the ravine woodland. In Chile, four species are restricted to scrub in a similar manner, and the fifth is restricted to woodland. The scrub habitat was further subdivided into nonrocky and rock substrates. The former is used by four species on both continents (three common and one rare in California; two common and two rare in Chile). The rock substrate supports, in California, three common species and two rare ones; and in Chile, again, three common species and but a single rare one. The distribution of lizard individuals among the three substrate types in sclerophyll scrub was closely parallel between continents, with proportions 0.52 and 0.48 (Chile and California, respectively) of individuals on the ground in shrubs, 0.40 and 0.43 on rocks in shrubs, and the remaining 0.08 and 0.09 on tree trunks in wooded ravines.

In the coastal scrub-type habitats of both continents, one finds three species, but in the Chilean species the individuals are rather more equitably distributed among substrate types (ground, rocks, and stems) than those of the Californian species. The Californian montane forest scores one extra species over the equivalent Chilean habitat (five vs. four), a difference that is largely offset by a striking size polymorphism in one of the Chilean lizards, *Liolaemus nigroviridis*. Although the distribution of individuals over four substrate types is similar in the montane forest (0.27 vs. 0.26, California vs. Chile, of the individuals in open ground sites; 0.08 vs. 0.25 on the ground under bushes; 0.50 vs. 0.39 on rocks; and 0.16 vs. 0.10 on trees), there is again a tendency for the Chilean species to be found more ubiquitously within a site than are the Californian species. Within-site diversities are thus very close, with equal α-diversities and perhaps a slightly greater point diversity in Chile.

The intercontinental similarities of lizard diversities extend throughout the changing season. At the height of the breeding season in both Chile and California, all five sclerophyll scrub species at the primary sites are common. Later in the season, when food supplies diminish, vegetation dries, and the days become hotter, the hatchlings of all species are active, but the adults of only one species are commonly found, both in California and in Chile. In both cases the species with active summer adults are the ant-eating lizards.

Species turnovers between sites on each continent can be evaluated to give β-diversities. All three species in the Californian coastal scrub are found in the sclerophyll scrub, and in the montane forest three more are gained while three are lost. Writing, as before, X for the rank of the next species gained at a certain point on the habitat axis, and Y for the rank of the species most recently lost at the same habitat position (see Figure 6-4), the Californian censuses can be represented as coordinates (X,Y) as follows: (3,0) in the coastal scrub, (5,0) in sclerophyll scrub, and (8,3) in montane woodland. The β-diversities between adjacent habitats are therefore $(5+0-3-0)/2(H_j - H_i) = 6/2(0.8) = 3.8$, or an overall β-diversity (for the gradient as estimated from the endpoint census locations) of $(8+3-3-0)/2(1.3) = 3.1$ species/per unit habitat shift. Coordinates representing the Chilean census are written (3,0) for coastal scrub, (6,1) for sclerophyll scrub, and (9,5) for montane woodland. These figures produce β-diversities of 4.0 and 5.0 between adjacent habitats and an overall β-diversity equal to 4.6 species/per unit habitat shift. Species turnovers with habitat change are therefore 1-1/2 times greater in Chile than in California.

Since we know that species total for California and Chile are respectively thirty-five and fifty-eight (although further fieldwork and systematic analysis might alter the latter figure), we can estimate γ-diversities as before, as the slopes of the species-area curves between 10^4 and 10^6 mi^2 ($\times$ 2.6 for km^2). For Chile, γ-diversity turns out to be 0.10; for California, 0.12. As before, the greater γ-diversity is found in the country where both species total and β-diversity are lower.

Birds

The birds of mediterranean California and Chile have been investigated in some detail, and the work already published (Cody, 1968, 1970, 1973, 1974, 1975) can be consulted for material and details beyond those summarized here. We reiterate that the studies deal with taxonomically distinct faunas in which the common and ecologically-analogous species are generally members of different families.

On each continent nine study sites were selected, to span the range of mediterranean habitats from coastal and valley grasslands through coastal scrub and sclerophyll scrub to evergreen woodland and coastal pine forest. Though some of these sites constituted matched pairs across continents, many only approximately matched their closest cross-continental analogs on the habitat gradient. But since a good number of sites were censused over a variety of habitats, α-and β-diversities are accurately calculated and compared as functions of habitat H.

A total of sixty-nine bird species was encountered in California, excluding aerial feeders, nest parasites, raptors, and nocturnal species. These species fell into fifty-one genera. In Chile only thirty-nine species were found at the census sites, over the same habitat range and sample, and the total includes only two pairs of congeners in thirty-seven genera.

Most of the species showed smooth distributions over the habitat gradient, and curves were fitted to represent numbers of species gained and numbers lost, $g(S)$ and $l(S)$, each as a function of habitat H. We can now compute diversity measures with simple, continuous function analogs of the measures used elsewhere, since the data are more complete and are fitted to curves, and any arbitrary interpoint comparison can be made on the gradient. The difference between the two fitted curves, $g(S) - l(S)$, gives α-diversity, and the derivation of the curve halfway between them, $d/dH\, g(S) + l(S)/2$, gives β-diversity (Cody, 1975). Each diversity measure is thus expressed as a function of position on the habitat gradient. These curves are shown in Figure 6-6.

In lower and more open habitats, up to and including chaparral at the primary sites ($H = \pm 1.0$), Chilean habitats show a slightly higher α-diversity. Beyond these sites, in taller chaparral, woodland, and forest, California sites show α-diversities increasingly greater than those of structurally similar Chilean habitats (Figure 6-6). At the extreme, mediterranean evergreen woodland (*Quercus agrifolia* in California and *Cryptocarya alba* in Chile) sup-

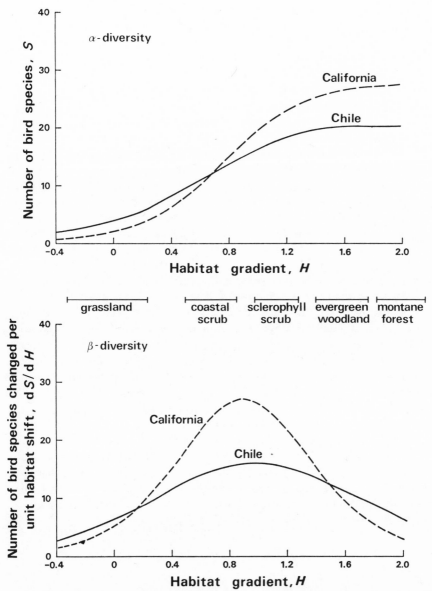

FIGURE 6-6. *Above, α-diversity for mediterranean birds plotted as a function of position on the habitat gradient* H. *Chile has rather larger α-diversity at low* H, *California greater α-diversity at higher* H. *In low chaparral and tall coastal scrub, the two continents are equal in α-diversity. Below, β-diversity is plotted as a function of position on habitat gradient* H; *California exceeds Chile in β-diversity over the larger central section of* H, *on which species turnover in California reaches levels almost twice as great as those reached in Chile (from Cody, 1975).*

ports nearly twice as many species in California (thirty-one) as it does in Chile (eighteen).

Species distributions were accurately assessed within habitats at several study sites to give point diversity. Within each of the chaparral primary sites, the Chilean species were more patchily distributed and yielded lower point diversities. Point diversities in grasslands were equal, but in coastal scrub the point diversity was again lower in Chile. In taller woodland habitats, which are relatively species-poor in Chile, most species occupied the whole study area there. But because of the wide discrepancy in species numbers, point diversities were again higher in California, even though species in the woodlands of the northern continent were point-specific within study sites.

Curves representing β-diversity are shown in Figure 6-6. As already indicated by the relative similarity in α-diversity, in conjunction with the much larger species count in California, Chile has a far lower β-diversity than does California. Species turnover per habitat shift peaks in intermediate habitats on both continents, around the sclerophyll scrub sites, but rather earlier in California than in Chile. Again, over a large section of the habitat gradient, which represents a habitat range from low scrub to tall chaparral, β-diversities in California are nearly double those of Chile.

Information on the higher diversity levels has already been given, in Figure 6-2. Beyond the β-diversity section of the curve, it is Chile that shows the greater γ-diversity and eventually gains species at a rate sufficient to bring total species numbers to virtual identity at the largest resolution of area. The similarity in these two endpoints is beguiling alongside the rather complicated differences in the species-area curves leading up to them. The overall picture is one of similar α-diversities; of point-diversity trends in the direction of "correcting" any discrepancy in α-diversity between continents (species-rich sites have more patchily distributed species within sites, thus reducing point diversity, whereas species-poor habitats support habitat generalists that tend to increase point diversity); and of β- and γ-diversity differences of large magnitude and opposite signs between continents but of different signs within continents. That the total species counts are so similar must be simply accidental.

Mammals

Each continent supports large mammal species, such as large browsers and carnivores, in its mediterranean habitats, but since the diversities and, particularly, the densities of these large mammals are extremely low, our studies were confined to mammal species smaller than cottontail rabbits. The sites studied include the primary sclerophyll scrub, montane forest, coastal scrub, and coastal succulent scrub sites on each continent, and almost all of the species encountered are rodents (Rodentia).

At the four California sites sampled, the numbers of mammal species present increases from five in succulent scrub to twelve in both the coastal scrub and the sclerophyll scrub sites and decreases to nine in the montane forest (Glanz, 1976). These are the same relative diversity trends shown by the insects studied, and also fit the lizard data; only bird diversity is a monotonically increasing function of increasing vegetation biomass and diversity as represented by increasing H. Considering only the common small-mammal species, peak species numbers occur in coastal scrub, and the four sites are characterized by, in California, three, eight, five, and three species, respectively. This species list has plainly identifiable desert, sclerophyll scrub, and montane elements, and the Camp Pendleton coastal scrub supports high numbers of both desert- and sclerophyll scrub-adapted species.

Chile, by contrast, has no such separate components in its small-mammal fauna, and many of the ten species encountered are of extremely broad distribution. Within sites, succulent scrub supports five to seven species, coastal scrub also seven species, sclerophyll scrub ten species, and the montane forest three species. Considering only common species, the counts increase from three to four from succulent scrub to both sage and sclerophyll scrub, and drop to a single common species in montane forest. Thus, the California sites have rather more common species than their Chilean equivalents, although trends of within-site diversity are similar along the parallel habitat gradients.

Since small-mammal species distinguish subhabitats within the study sites, such as ravines vs. hillside slopes, and rocky vs. shrubby slopes, the α-diversities lie somewhere between site totals and the totals for each trapping location within a site. At the primary site in California, from two to four common species and from two to nine (mode five) total species were caught at each of the nineteen 0.5 ha trap locations; in Chile, two to six common species and from one to ten (mode five) total species were caught in a similar set of nineteen trap locations. At smaller resolutions of area, at the 0.1-ha level, small-mammal densities are lower in Chile but species number is higher (mode four, vs. three in California); Chile also produces more small-mammal species per individual trap location, and hence has higher point diversity than California. Thus, Chile achieves species diversity equal to California's with lower numbers of individuals and lower densities. As sample size decreases, in terms of numbers of individuals caught and total area sampled, small-mammal diversity drops off at a greater rate in California than in Chile. In sum, α-diversity is equal at the resolution of 0.5 ha; but as sample size decreases, diversity per individual is somewhat higher in Chile, and diversity per unit area is less unequivocally higher in Chile.

In β-diversity the small-mammal picture is clearer. The lowest species count in California has species coordinates (X, Y) of $(4,0)$ at $H = 0.3$ and of $(18,9)$ at the taller end of the gradient, Mount Laguna, where $H = 1.9$. Overall β-diversity is therefore $(18+9-4-0)/2(1.9-0.3) = 7.2$ species/per unit habitat shift. In Chile the coordinates at the extremes of the habitat gradient are

(5,0) at H = 0.3 and (10,7) at H = 1.8. Overall β-diversity is therefore 4.0 species/per unit habitat shift, only just over one-half that in California. On gradients into the drier desert habitats of the Norte Chico in Chile and Baja California in North America, the difference in β-diversity between continents becomes even greater. Chile shows only a gradual species loss (down to a single species), without any species replacement or supplementation, whereas the move to Baja California produces a rapid species turnover (from samples of four additional sites in Chile, two in Baja California; Glanz, 1976, Meserve and Glanz, 1977).

Again, assuming that species turnover among mediterranean habitats is similar to species turnover among habitats in general, such that the β-diversity trend typifies areal samples up to 10^4 mi^2 (2.6×10^4 km^2), we can deduce estimates of γ-diversity. Knowing the total number of small-mammal species for the central provinces of Chile to be around twenty-six, and that for the country as a whole to be thirty-seven (Osgood, 1943) to forty, since new records in peripheral areas are increasing the number, the γ-diversity for Chile proves to be 0.15, the slope of the last section of the species-area curve for the country. California has eighty-five small-mammal species, over twice the number in Chile, but since most of the diversity is accounted for by local differences between habitats, γ-diversity for California is calculated at only 0.06. Thus, Chilean γ-diversity exceeds that of California by a factor of 2.5. Just as for birds and ants, Chile has the lower β-diversities and species totals, but the higher γ-diversities.

Summary of Diversity and Distribution Patterns

We selected a series of matched study sites in the mediterranean regions of Chile and California, and compared species numbers for certain select groups of consumers within sites, both within and between continents. We then compared between continents the rates of species turnover between habitats on each continent. We also estimated the rates of species addition with increasing sample area on each continent, to the extent this was possible, independently of habitat changes with increasing area.

Intercontinental comparisons involved species with very different taxonomic affiliations, especially among the vertebrates, and the comparisons between analogous species on different continents involved representatives of different families.

Of all intercontinental comparisons of species diversities, the greatest similarities were always in α-diversity, the number of species of a particular consumer taxon occurring in an apparently homogeneous habitat patch. We found that α-diversities are virtually identical in lizards, birds, and mammals, but up to 1–1/2 times greater in bees and up to 2 times greater in ants in California. We consider these similarities in α-diversity to constitute strong evidence that competition acts in a similar way on both continents to regulate the species-packing levels of consumers in analogous communities, and to follow directly from competition theory of consumer use of parallel resource gradients.

We rank habitats on a gradient H that reflects changing environmental conditions, most prominently available moisture. This ordination is intended to ensure that habitats that are similarly viewed by the organisms studied, both consumers and plants, fall in adjacent sectors of H; this results, if it is a good generalization, in smooth, contiguous species distributions over H. In birds, α-diversity increases continuously with increasing H; but for bees, lizards, mammals, and ants, α-diversity peaks in the chaparral (just toward the higher H side of the primary sites in lizards, just toward the lower H side of the primary sites in mammals). Using species turnovers along H, we find that β-diversity is two to three times higher in California than in Chile for those consumer groups, including ants, bees, birds, and mammals, that have higher total species numbers in California. Lizards, with higher total species numbers in Chile, accordingly have higher β-diversity in Chile.

Here we can recall the patterns of diversity of the plants themselves, relative to our habitat gradient H. Chilean plants are higher in α-diversity than Californian plants, and show less species turnover, or β-diversity, with shift along the moisture gradient, than their Californian counterparts. Since the gradient H is based chiefly on plant structure, could the plant-diversity differences between Chile and California account directly for any of the consumer-diversity differences? The answer is almost certainly "no," since in those consumer taxa least dependent on plant-species diversity, the birds and mammals, the trends in plant vs. consumer diversity are parallel in the two countries (high α-diversity, low β-diversity occur simultaneously in both plants and consumers). Yet, in the consumer taxon that we would expect to be most closely tied to plant-species diversity, the insects, high plant diversity is found where insect diversity is lowest.

We assume that central sections of the species-area curves used here reflect β-diversity, and that the final sections of the curves reflect γ-diversity. The assumption is somewhat crude, since for some organisms for which habitats appear to be quite uniform and extensive, or for which climate per se produces the closest estimate of "habitat" structure, geographic replacements may be boosting species accumulation with increased sampling areas *before* species turnovers follow habitat changes; i.e., γ-diversity at some point precedes increased β-diversity with increasing sample area. Here we generally lack the expertise to make such distinctions.

In estimating γ-diversities from the slopes of the terminal sections of the species-area curves, we observe a consistent relationship to β-diversity: where the latter is higher in California than in Chile for a particular consumer group, γ-diversity is lower in California; similarly, for lizards, β-diversity is higher in Chile, γ-diversity higher in California. Thus, the two measures appear to be inversely related.

Although competition regulates the structure and organization of local communities, it has less influence on such variables as species distribution and range, or on total species numbers within larger geographic regions. We therefore suspect that β- and γ-diversity are strongly affected by the accessibility of various habitats to various groups of species that differ in their mobility and dispersal capacities. The intrusion of variously complete or effective barriers between habitats or geographic regions, and the existence of factors that affect speciation rates, adaptive-radiation potentials, and the dispersal of species through variously inhospitable topography and geography, must all exert an influence on patterns of β- and γ-diversity.

PATTERNS OF DENSITY AND RESOURCE USE

Habitat as a Resource Gradient

Most of the theory of species packing concerns the accommodation of species on resources (such as food) that are consumed at certain rates with certain efficiencies but also renew at certain rates and thus maintain themselves at certain levels. This species-packing concept affects chiefly point diversity and α-diversity, and we shall discuss these effects in this section. But first, we shall do well to consider another type of species packing, one that concerns the accommodation and distribution of species over habitat gradients, and therefore affects β-diversity.

Sets of habitats, such as those of the mediterranean regions that range from grassland to evergreen woodland, constitute a rather different sort of resource gradient; species are arranged according to competitor pressure and the performance of each species in each habitat type. Competition does not fit a combination of utilization curves to a resource-production curve; rather, species expand or restrict their use of habitats so that the set of species forming the community at any point on the habitat gradient H conforms to the limits of α-diversity set by the diversity of food types and/or foraging sites at H.

The relative constancy of α-diversity in our results indicates that a good deal of adjustment in the habitat distributions of species has taken place across time. Specifically, species-rich habitat gradients must show higher β-diversity with smaller habitat niche breadths per species than species-poor habitat gradients shown, as long as α-diversity is constant. For minor variations, low α-diversity can be partially compensated by niche expansion in the habitat dimension, whereas high α-diversity restricts such expansion. Thus, for Californian birds over the low range of H, where species packing is relatively looser than in Chile, the mean habitat niche breadth B_H for all species occurring over low H is 0.55; for the Chilean birds it is less, 0.35 H units (Cody, 1974). Over both intermediate and high ranges of H, California shows tighter species packing than Chile, particularly for H greater than 1.4. The mean B_H values for intermediate and high H, averaged over all species found there, are 0.41 and 0.46, respectively. In Chile, over these same habitat ranges, B_H values are increased, to 0.54 and 0.75, respectively (Cody, 1975). Habitat use is thus moderately expanded where α-diversities are slightly lower, but virtually doubled where exceptionally low α-diversities permit extensive habitat expansion.

Resources and Trophic Relations

Though the organisms discussed in this chapter have been considered collectively as "consumers," all they have in common as a group is the fact that they cannot photosynthesize and must ultimately depend on green plants for their sustenance. Between them, they span a variety of trophic levels, and are involved in a much larger web of trophic relations that we have only partially investigated. Our intention in this study was not, however, to dissect and document trophic relations. Rather we sought, by choosing certain select and discrete groups of taxonomically related consumers, to substantiate the concept of convergent evolution at the species and supraspecies levels. After only the briefest consideration, one realizes that such substantiation is more likely to be possible—and where possible more likely to be successful—in those groups of consumers whose trophic relations are not tangled with other consumers in other taxa. Still, where trophic relations do exist across group boundaries, they can be elucidated and compensated by studies of both or all of the consumer groups involved; such studies are more difficult where trophic relations occur in parallel (within a trophic level), less difficult where they occur in series (between trophic levels).

The consumers we chose to study in detail are therefore the more trophically discrete groups of taxa, in the sense that competition for shared resources is stronger within those groups than between groups of more varied taxa. Figure 6-7 affords a rough picture of the trophic relations in mediterranean scrub, and includes the major groups of consumers. Four major resource types are produced from plants: foliage and juices, nectar and pollen, seeds, and leaf litter. Of the consumers that feed directly on plant resources, we studied in detail the pollinator insects, measured the biomass and activity of flying insects in general, and collected considerable data on soil and leaf-litter insects (Saiz, unpublished, for the last category). In addition, some small mammals, birds, ants, and lizards feed directly on plant materials. The herbivorous insects and their predators provide the bulk of the resources for insectivorous birds and lizards, as well as for many ants and mammals, and it is in this second stage of consumer interactions that our studies were concentrated. Consumers here are themselves the resources for the top predators, chiefly mammals and birds; but we have almost totally ignored this third level of consumer-resource links.

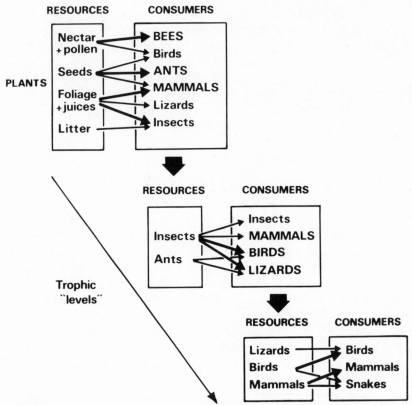

FIGURE 6-7. *Approximate trophic relations of the taxa in the communities studied. Major pathways of energy transfer are indicated by thicker arrows. Capitalized consumers are those we studied. See text for discussion.*

Note that this rough scheme very emphatically does not provide a complete picture of scrub trophic relations, merely an approximate guide to the consumer relations we actually studied; the true food web is vastly more involved and reticulate. In this representation (Figure 6-7) we indicate (by heavy arrows and type) those consumers and interactions that most occupied us, and (by lighter arrows and type) those of a more peripheral interest (despite their importance to the community and their perhaps overwhelming proportion of the total animal biomass of the scrub habitat).

The diagram in Figure 6-7 shows that most of the birds eat insects, as do most of the lizards. But a few birds eat nectar and fruits, and others eat ants, lizards, or other birds. Likewise, a few lizards are herbivorous, and a few eat ants. Mammals eat mostly seeds, foliage, or insects, and are eaten by raptorial birds or by larger mammal predators. Ants eat seeds (conspicuously only in California) and insects, and themselves fall prey to other ants, to one very specialized Californian lizard (*Phrynosoma*), and to less specialized lizards and birds on both continents.

Resources and Utilization Curves

Earlier it was shown (Figure 6-1 and text) that resource use in ecological-ly-related species can be concisely illustrated by showing the distribution of species over resource gradients as series of utilization curves. When consumers are depicted in this way, parallels can easily be seen, their accuracy assessed, and further predictions made. Figure 6-8 shows the lizard communities for the Californian and Chilean montane forest communities. There are five species in California (upper histograms) and four in Chile (lower histograms). But in one Chilean species (*Liolaemus nigroviridis*) an unusually large number of size and age classes are active simultaneously, and the species can be divided into two "ecospecies" on the basis of habitat use, body size, and food habits, one consisting of the small individuals, the other of the larger individuals (Fuentes, 1974, 1975).

In passing we should mention that the sex dimorphism and the coexistence of several age classes that we see in these lizards seem to be rather common mechanisms in species exposed to selection for increased niche breadth in the absence of competing species (see, for example, Roughgarden, 1972). Thus, *Lacerta tiliguerta,* which lives in Sardinia where island effect is notable, exhibits greater size polymorphism than its close congener *L. muralis,* which lives in sympatry with larger congeners in southern France (Fuentes, unpubl.).

The histograms of Figure 6-8 show utilization for each species in each of five food categories: from the top down, small, medium, and large non-ant arthropods (1-3), ants (4), and, at the bottom, vegetation (5). The insects were sampled at ground level in this study site using Tanglefoot plates and can traps; estimates of the frequencies of arthropod resources in the first four categories based on this sampling are given in the lower part of the figure.

The species within each of the columns from A to E are considered ecological analogs, on the basis of microhabitat preference and relative body size. The species in columns A and C forage on the ground under bushes, and differ between columns largely in the size of the food they capture. The mean head lengths of the two Californian lizard species are 9.9 mm (*Eumeces*) and 23.0 mm (*Gerrhonotus*); of the two Chilean species, 13.3 mm (*Liolaemus schroederi*) and 24.0 mm (*Urostrophus*). All four species are secretive and rare (relative frequencies of small and large species in California, 0.07 and 0.01, respectively; in Chile, 0.15 and 0.01, respectively). Correspondence is good between the diets of the two smaller species analogs (overlap 70 percent), less good between the diets of the two larger species analogs (overlap 60 percent).

Each continent has a species that forages over open ground: in California this is *Uta* (head length 11.5 mm); in Chile it is the juvenile individuals of *Liolaemus nigroviridis* (head length 8-13 mm, average 11.5 mm). These two species exhibit very similar utilization curves, both feeding mostly on small insects, to a lesser extent on medium-sized insects and ants (which are tallied separately). In addition, though small *L. nigroviridis* eat some vegetation, the two utilization curves nonetheless overlap at the 86 percent level. Each con-

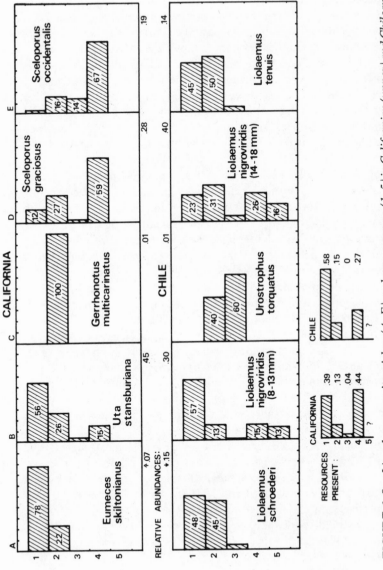

FIGURE 6–8. *Lizard community niches (A–E) and resource use (1–5) in Californian (upper) and Chilean (lower) montane forest. Corresponding niches are characterized by species on both continents that are similar in body size and proportions, feeding behavior and diet (see histograms), density, and habitat preference. See text for discussion (from Fuentes, 1974).*

tinent also has a rock-dwelling lizard, *Sceloporus graciosus* in California (head length 11.2 mm) and the large *L. nigroviridis* (head length centered at 16.0 mm) in Chile. Their food preferences coincide at medium-sized insects; and their utilization curves overlap 67 percent, the Californian species eating relatively more ants, its Chilean analog again eating some vegetation. Finally, each continent has a tree-dwelling lizard, *Sceloporus occidentalis* in California (head length 15.0 mm) and *Liolaemus tenuis* in Chile (head length 12.4 mm), with utilization curves overlapping at a much lower 33 percent (for *S. occidentalis* and *L. tenuis*).

Within each community the summed utilization curves of the lizards bear some resemblance to the production curves for their food resources. Thus, fewer small insects were captured on the Tanglefoot plates in California than in Chile, and the utilization of the small-insect category is heavier in Chile than in California. Similarly, many more ants were caught in California than in Chile (44 percent vs. 27 percent of the total arthropod catch, a result that parallels the α-diversities of the ants for these two study sites); correspondingly, in California there are two lizard species (*Sceloporus*) that depend heavily on ants in their diets, and one less frequent ant user (*Uta*), whereas the Chilean species use ants only lightly. The use of vegetation by the Chilean *L. nigroviridis* remains anomalous, but does correlate with a greater density of low herbs at the site in Chile (see Table 5-5, and the earlier section on pollinators).

From Figure 6-8 we can draw additional conclusions on the densities of particular species, but in the absence of data on absolute densities these are difficult to verify. From the production curves we would expect the lizards of column C to be rare, particularly *Urostrophus*. Since the absolute numbers of insects caught were considerably greater in Chile than in California, the species of columns A and B should be more common in Chile than in California. And since ants were both relatively and absolutely more common in California, the lizard species of columns D and E should be commoner in California than in Chile. Our qualitative observations on the lizards bear out these projections, but the problems of sampling lizard density have so far prevented definitive tests and verification.

Utilization curves such as those shown in Figure 6-8 are but one aspect of the comparative biology of the consumers they represent. When these are considered in conjunction with body size, body proportion, and behavioral similarities, the lizards that occupy similar microhabitats are found to be similar in most other aspects of their general biology. Such discrepancies as do occur are, in general, readily interpreted through differences in the resource base that supports the community.

Similar qualitative statements can be derived for insectivorous birds, which exploit a generally different segment of the insect population—mobile insects that are resting or feeding on the foliage well above the ground, or else those traveling through the air. The densities of such insects were estimated in several ways, including the use of Malaise traps; the seasonal distrib-

ution of insect abundance at the primary sites from Malaise trap catches is
shown in Figure 6-9 (M. Atkins, unpubl.).

Although there is an approximate similarity in the dry weight of the
insects caught in California and Chile, there are some obvious differences.
Most notably, there is a higher spring peak of insect abundance in California,
lower winter levels of insects in California, and a secondary fall peak of insect
production in Chile. These differences can be interpreted in the light of the
data in chapter 5 on plant seasonality and production. The Californian
chaparral (sclerophyll scrub) has a much more marked early-season peak in
flowering activity and a sharper peak in fruiting season than the equivalent
Chilean sclerophyll scrub (matorral), and the matorral species as a group have
much more extended reproduction seasons (Figure 5-20). Thus, plots of the
percentage of ages of plant species flowering and plant species fruiting sub-
tend a considerably greater area in Chile (by a factor of 2.5) than in California, even though the relation between flowering season and fruiting season in
the two countries is the same, with the latter 1.3 times the length of the
former in both cases.

In Figure 6-10 we plot the cumulative area under the curve of insect biomass of Figure 6-9 against cumulative area under the combined curves of
percent species flowering and fruiting (Figure 5-20). The areas under the two

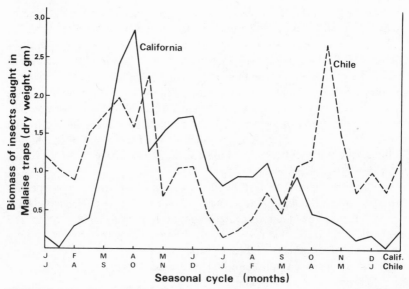

FIGURE 6-9. *Seasonal distribution of insect biomass caught in Malaise traps
erected at the primary sites in Chile and California. Note that although California has a higher spring peak, only Chile has a fall peak; and that Chile has,
as well, higher winter levels of insects. See text for discussion (M. Atkins,
unpubl.; data from single 12-month study intervals).*

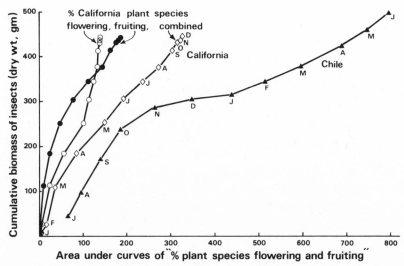

FIGURE 6-10. *Relationship between insect abundance (the Malaise trap data of Figure 6-9) and percentages of plant species in the flora that are flowering and fruiting, in both California (three curves at upper left) and Chile (curve at lower right). See text for discussion.*

insect-biomass curves are similar, but the Chilean total is 12 percent greater (497 units vs. 442) and accumulated at a far slower rate overall, and the California spring production (mid-March to the end of May) is greater by 20 percent. Insect biomass in California's early season (January through early May) is about double the level of that in Chile's early season, but the rate of accumulation of insect biomass during this period with increasing area under the percentage-flowering and percentage-fruiting curves is the same. This rate of accumulation of insect biomass continues uniformly through the year in California (constant slope of curve in Figure 6-10), until the summed areas under the plant curves reach a December endpoint of 323 units (140 under the percentage-flowering curve, 183 under the percentage-fruiting curve). But in Chile there is a mid season lull, owing perhaps to the adverse conditions for both plant photosynthesis and insect activity in the hottest and driest part of the year, after which a renewed fall accumulation of insect biomass continues until the June endpoint reaches a value of 798 units under the plant curves (348 under the percentage-flowering curve, 450 under the percentage-fruiting curve).

To this picture we can add the information on shrub biomass. At the California site, overall plant biomass is three times that at the Chilean site; also, shoot production is greater in California by a factor of 2.8 and leaf production greater by a factor of 2.0 (see Chapter 5). These data help to explain the higher insect densities in California relative to degree of plant reproductive activity. In summary, more vegetation cover and more plant biomass produce more insects, but extended seasons of plant production, especially in the flowering and fruiting seasons, favor higher insect abundance over longer periods of the year, especially in late summer and fall.

These similarities and differences in mobile-insect production are translated into similarities and differences in bird densities as follows. The combined abundance of the six main species of insectivorous birds in California totals 6.38 pairs per ha, 10 percent greater than the abundance of their six ecological counterparts in Chile (Cody, 1973). But these counts were made in spring at the height of the breeding season, and thus parallel the greater spring insect production in California. Overall bird density in Chile is in fact 16.25 pr/ha, 16 percent greater than the California total of 14.0 pr/ha. This difference might reflect the greater year-round abundance of Chilean insects, but more likely is owing to the greater density of ground-foraging (insectivorous) birds in Chile. There are nine species in Chile (vs. five in California), totaling 7 pr/ha (vs. 5.25 pr/ha in California). This major discrepancy, which parallels the trend for higher densities of small insectivorous ground lizards in Chile, is due to (1) the more open, more patchy nature of the Chilean matorral (see Chapter 5), which provides a greater diversity of bird-foraging sites that in turn support additional bird species, and (2) the greater abundance of ground-level insects in Chile, as measured by the Tanglefoot plates. The greater abundance of gound-level insects is in turn correlated with the greater production of annuals and other lower-stratum herbs in Chile, especially in the mature matorral, the plants of which have no counterparts in mature California chaparral. Moreover, these herbs produce the seeds that support certain insects and adult finches with insect-eating young (e.g., *Zonotrichia capensis*), and again the broken character of the vegetation structure in Chile adds a diversity of microhabitats that favors higher insect diversity and abundance.

The open and patchy nature of the Chilean matorral has another effect on the insectivorous bird community. Since any particular height interval of the chaparral vegetation, e.g., that between 2 and 3 m off the ground, is less constant and less predictable over space in Chile than in California, the six main Chilean insectivores must markedly increase their niche breadths over vegetation height—i.e., become more generalized in vertical foraging zones. Their niche breadths average 3.25 vegetation "layers," vs. an average 2.55 for those in California. But with the incorporation of extra ground-foraging species with very similar vertical foraging zones, the overall picture of community organization with respect to niche overlaps is strikingly consistent: within each primary site, niche overlap in habitat use is 72 percent in California, 75 percent in Chile; in food/feeding behavior, 48 percent in California, 51 percent in Chile; in vertical-foraging zonation, 27 percent in California, 26 percent in Chile.

A final comment on the exceptional fall insect production in Chile is in order. Though this potential insect food does not appear to be used by breeding birds, we can predict that it should influence the Chilean matorral species in (1) longer breeding seasons, (2) more temporal segregation among species in breeding season, (3) more broods per year, and (4) a reduced migratory tendency and increased tendency to year-round residency. Only the last of these influences can be surmised at the present, and our expectations are fulfilled: whereas over 30 percent of the bird species of the California chaparral are migratory (five or six of seventeen species), just two of nineteen Chilean matorral species (the giant hummingbird *Patagonas gigas* and the white-crowned elaenia *Elania albiceps*) migrate.

Specialization, Generalization, and Niche Overlap

Competitor communities may attain high diversity in various ways—through a high degree of specialization of certain member species that come to monopolize a specific, limited range of resources, through a high degree of niche overlap among more generalized sets of species, or through some combination of these two extremes. The food-web relations between flowering plants and flower-feeding herbivores accommodate a wide range of possible combinations of specialists and generalists within the same community. Our studies of the pollinator faunas of Chile and California can illustrate some of these possibilities.

The floral resources at each site were quantified by detailed censuses of 10 m X 10 m plots of vegetation, with results as shown in Table 6-1. Since the Californian chaparral system is comprised of two segments (the post-fire annuals and geophytes and the shrubby perennials to which they progressively give way after a recovery period of several years), data were collected separately for these temporally-distinct phases; in Chile (see Chapter 5) there is no such temporal segregation of the herbaceous and the woody vegetation.

The Chilean matorral plots contain 108 plant species, similar to the total of 103 from burned and nonburned chaparral plots. In addition, the total number of woody plant species is similar in Chilean matorral (about 246: 217 trees and shrubs, 29 succulents; Zollner, pers. comm.) as in chaparral in California (about 249: 217 trees and shrubs, 32 succulents; Munz and Keck, 1955). But at any one time in a Californian plot, plant species numbers are lower than in a comparable Chilean plot, and larger censuses (of 1,000 m^2) produced thirty Chilean shrub species compared to nineteen in California. The floral biomass in California is somewhat more comparable to that in Chile, but the floral biomass diversity is still considerably reduced relative to the matorral figure (see Table 6-1).

If the reduced diversity in local plant censuses in California is due to competitive colonization after fires, we might expect a similarly reduced diversity in the Californian floral herbivores. Table 6-1 shows this is not the case, for nearly 400 species were found in combined Californian chaparral plots, versus just 281 in Chilean matorral. This disparity between the two continents is

TABLE 6-1 *Abundance and Diversity of Plants and Pollinators*

Site	Plants in census plot:[a] # Individuals	# Species	Plant species diversity H	Floral biomass[b] F	F diversity	# Pollinator pollinator taxa[c]	# Pollinator biomass P	Biomass ratio P/F
California								
Coastal Scrub	79,889	83	2.15	2,755	2.82	199	644	0.24
Chaparral	2,636	44 } 103[d]	2.79	8,969	2.11	309 } 388[d]	3012	0.33
Chaparral (burn)	43,236	85	2.89	2,625	2.46	254	2236	0.88
Desert scrub	41,110	39	1.66	2,200	1.53	182	1530	0.70
Chile								
Coastal scrub	110,055	131	2.46	16,508	2.71	130	946	0.06
Matorral	270,853	108	1.12	10,302	2.84	281	2549	0.24
Desert scrub	79,019	76	1.03	2,508	2.71	100	700	0.28

a. Census plots are 10m × 10m.
b. Floral biomass data taken from 1,000 m² plots.
c. Pollinator data taken from 500 m² plots.
d. Figures are numbers of distinct taxa in combined burned and unburned chaparral sites.

178

duplicated at the other sites, and the ratio of flower-herbivore species to plant species is two to four times higher in California than in Chile.

Although the California site supports twice the number of species (and twice the number of individuals) of floral herbivores, the ratios of pollinator biomass to floral resource biomass are much more similar (Table 6-1), indicating the presence in California of many small-bodied extremely abundant bee and beetle species. These species are generalized opportunistic feeders for the most part, though enormous numbers of some are associated with only one or two related plant species (e.g., *Perdita rhois* (Andrenidae) on *Adenostoma fasciculatum* and *Phthiria* spp. (Bombyliidae) on *Ceanothus leucodermis*). Most of these tiny flower-herbivore species are of little, if any, service to the plants as pollinators; they do act to consume the resource and force the more useful pollinators to forage more widely and hence increase the efficiency of cross-pollination for these taxa.

Of much greater significance in the mediterranean scrub communities in both continents are the rather small (length approximately 5-10 mm) halictine bees. Details of their life-history are very poorly known, but most are generalist, opportunistic feeders which are either multivoltine (several generations per year) or long-lived, semi-social or eusocial, colonial species. The deep-rooted water-storing capacity of many chaparral and matorral shrubs permits an extended blooming season of nonoverlapping blooming peaks for the component species (Mooney, 1972); these dominant plants are utilized in succession by these very abundant halictine bees, which serve then as very efficient pollinators due to relatively nonoverlapping anthesis periods in the dominant plants.

In both mediterranean scrub communities, feeding-specialist bees are both abundant and diverse. Such specialists are associated with shrubby, sub-shrubby, and ephemeral elements in the ecosystem. These specialists may be genetically programmed to visit only one or two closely related species of plant resources, regardless of patterns of local distribution, or may be opportunistic feeders that are forced by economical considerations of foraging efficiency to specialize on only one local resource. It is often difficult to distinguish between these two foraging strategies operationally, but attempts were made to observe the flower-visiting habits of species resident in the research sites, as well as in other sites of different resource sets and different competitive regimes. For Chilean bee species the consistency and purity of pollen loads was examined through the microscope and in California similar studies were supplemented by analyses of floral records and pollen loads in museum collections (Moldenke and Neff, 1974; Moldenke, 1976; Moldenke and Toro, in prep.). Examples of the most important specialist pollinator taxa in two habitat types are presented in Table 6-2.

The final guild of pollinators present in mediterranean scrub communities are the large, fast-flying, high-energy-requiring long-lived species which exploit the changing array of the very "best" resources (in terms of quantity

TABLE 6-2 *Example of Specialist Pollinator Taxa in Two Habitat Types in Chile and California*

	CALIFORNIA		CHILE	
	Plant species	Pollinators	Plant species	Pollinators
Coastal Scrub				
	Camissonia spp.	Andrena anatolis, A. oenotherae, A. rozeni, Hesperapis nitidula	Adesmia, Astragalus Astragalus canescens Calceolaria nudicaulis	Anthidium gayi Alloscritetica tristrigata Tapinotaspis caerulea, Centris nigerrima
	Ceanothus verrucosus	Andrena candida		
	Encelia, Coreopsis	Xenoglossodes davidsoni	Sphaeralcea spp.	Oediscelis sp. Spinoliella maculata, S. herbsti
	Lasthenia chrysostoma	Andrena baeriae		
	Opuntia, Mammilaria	Ashmeadiella opuntiae, Diadasia australis, D. rinconis	Trichocereus chilensis	Trichoturgus dubius
Chaparral				
	Calochortus splendens	Perdita califonica	Adesmia angustifolia	Allanthidium rodolphi, Panurginus sp. Psaenythia parvula Anthidium colliguayanum, A. penai
	Camissonia leptocarpa	Andrena blaisdelli		
	Cirsium californicum	Osmia californica		
	Cordylanthus filifolius	Chalicodoma angelarum	Adesmia arborea	Centris cineraria, nigerrima, Tapinotaspis herbsti
	Cryptantha spp.	Andrena cryptanthae, Proteriades jacintana, P. nanula, P. semirufa, P. tristis	Calceolaria spp.	
	Convolvulus aridus	Diadasia bituberculata	Cassia klausiana	Leioproctus zonalis
	Cryptantha, Emmenanthe	Chelostoma californicum	Colletia spinosa	Leioproctus fazii
	Cryptantha, Phacelia	Conanthalictus nigricans	Loasa tricolor	Leioproctus rufiventris, Oediscelis sp.
	Eriastrum sappihrinum	Perdita rivalis		

180

TABLE 6-2 (Continued)

CALIFORNIA		CHILE	
Plant species	Pollinators	Plant species	Pollinators
		Oxalis laxa	Panurginus nigroaeneus
		Psoralea glandulosa	Notanthidium steloides
		Senecio, Haplopappus, Chaetanthera	Diadasia chilensis
		Stachys grandidentata	Oediscelis sp.
		Trichocereus chilensis	Leioproctus semicyaneus
Eriodictyon angustifolium	Nomadopsis linsleyi		
Eschscholzia californica	Nomadopsis obscurella		
Lasthenia chrysostoma	Andrena puthua, A. submoesta		
Penstemon spectabilis	Anthocopa triodonta		
Phacelia spp.	Chelostoma bernardinum, Conanthalictus macrops, Dufourea cuprea, D. mulleri, Osmia bruneri		
Phacelia, Emmenanthe	Chelostoma minutum, Conanthalictus namatophilus		
Rhamnus, Ceanothus	Perdita micheneri		
Trichostema parishii	Ashmeadiella salviae		
Adenostoma fasciculatum	Hesperapis ilicifoliae, Perdita fieldi		
Arctostaphylos glauca	Andrena arctostaphyllae, A. vandykei		
Ceanothus spp.	Andrena angustitarsata, A. atypica, A. candida, A. ceanothifloris, Panurginus gabrielis		
Ceanothus, Rhamnus	Andrena chlorogaster		
Eriogonum fasciculatum	Perdita claypolei		
Gutierrezia, Haplopappus	Melissodes lupina		
Lotus scoparius	Hoplitis howardi, H. producta bernardina		
Malacothamnus densiflorus	Diadasia laticauda, D. nitidifrons		

and caloric content of nectar, i.e., Watt et al., 1974). This guild is composed of hummingbirds, sphinx moths, and large heavy-bodied bees. Though present in both continental sites, this guild does not assume a major importance, and in both countries only 10 percent of the resident plant species rely upon its services for pollination (Moldenke, in prep.).

In the chaparral of both continents flower-herbivores are diverse and abundant, with most individuals generalized and most species specialized on an abundant and varied resource base. In contrast, the environment of the coastal scrub is foggy, cooler, and more temperate, and climatically much less ideal for most pollinator groups. The ratio of total pollinator biomass to total nonanemophilous floral biomass decreases to the lowest levels observed in these studies. All pollinator taxa decrease in diversity and most small-bodied species become quite rare or are constrained to activity periods encompassing only the warmest parts of the day. Large-bodied thermoregulatory species become the most conspicuous guild of the community, as their activity is not affected during the frequent days of heavy fog. Generalist feeders are the most abundant and diverse of all pollinators, as the competition for limiting resources in the mediterranean scrub switches to a system of plants competing for limiting pollinators, which are restricted in the distances they can fly by the coolness of the weather.

As aridity becomes more pronounced along the transition to the desert sites in both continents, community floral biomass drops by 80 percent as intershrub distance increases and annual plant density decreases. Both plant species count and floral-herbivore species count decrease by 50 percent from the levels found in mediterranean scrub. Annual variability, however, is the most critical aspect of desert pollination systems. During years of diminished (but not zero) rainfall, plant response is minimal and floral visitors are concentrated from surrounding areas resulting in heavy competition by vectors for plant resources. In contrast, during years of heavy rainfall both shrub and annual plant response is very heavy and the same cool windy climate propitious for the plants severely inhibits pollinator activity and competition for pollinators leaves the great majority of plants unexploited.

Since the resources for desert pollinators are presented in alternating phases of dearth and superabundance, exploitation patterns have evolved for both circumstances. In years of superabundance both generalist and specialist feeding species are easily supported, but in low-rainfall years specialist-feeders, which are morphologically adapted to exploiting specific resources and are genetically selected for synchronous emergency with their resources, are at a strong selective advantage. In such years generalist feeders subsist on the minute resources offered by desert annuals or the shrubs heavily attended by specialists; the colonial social bee strategy so prevalent in other ecosystems is precluded. Specialist-feeding species, therefore, are the rule in the desert ecosystems, comprising the largest number of taxa and by far the largest number of individuals.

In summary, there are extrinsic factors on pollinator communities that affect intrinsic organization in direct and indirect ways. Since pollinator taxa belong to a wide variety of unrelated taxonomic groups, one would expect that, along an environmental gradient in which many features of environment are changing, different taxonomic groups will be limited differentially by environmental features other than food competition. Physiological traits such as tolerance of cold, or aridity, may be critically important in different ways to different groups. In the deserts of California and Chile small solitary bees, bombyliid flies, and wasps predominate. Along the cooler coastal margins large thermoregulatory bees, long-tongued flies, and hummingbirds are predominant. In the evergreen scrub all groups do well, in addition to butterflies and hoverflies. Such faunal changes do not necessarily imply changes in food utilization patterns, since within each taxonomic group of pollinators all possibilities occur between extreme generalization and extreme diet specificity.

Extrinsic factors affect resource utilization by pollinators in similar ways on the two continents. The following trends are evident in both California and Chile: (a) emphasis on feeding generalists by species and individuals in the mediterranean scrub communities; (b) emphasis on feeding generalists by species and biomass in the more coastal communities; (c) emphasis on feeding specialists as virtually the only adaptive lifestyle in the deserts; (d) in addition, there is complete dominance of community food webs by very few species of supergeneralist feeders in the forests and in coastal regions of higher latitudes.

Niche Correspondence at the Species Level

We face now the question whether natural selection has pushed convergent evolution to replicate on separate continents the morphological and ecological attributes of species occupying the same niches. The answer, not a simple one, varies both among consumer groups and within the same consumer group. As before, then, we shall treat consumer groups consecutively and attempt to explain the variability in results.

In neither pollinator insects nor in ants are there striking one-to-one correspondences at the species level. The numbers of species involved are large, and each species appears to be a unique solution to particular details of microenvironments and substrates and to particular intricacies of plant structures and chemistries. These details and intricacies, firstly, were not assured a cross-continental match by the relatively crude climate and habitat parallels used to select study sites; moreover, they involve a set of plant-consumer interrelations that ecologists have, in general, scarcely begun to unravel. In addition, these lower organisms, with their primitive, stereotyped behavior, may be less likely to find efficient or optimal solutions to the problems of resource exploitation and predator avoidance, and must therefore depend heavily on their genetic heritage and an adaptive repertoire that either copes in the time-honored fashion or sees its owner eliminated before we could witness its performance. Any similarity among quasi-analogous ecospecies in these lower groups seems much more readily attributable to common ancestry than to convergence to a common phenotype from different genetic and morphological origins, and the observed convergences are those between analogous multispecies sets of consumers rather than between single species counterparts.

Lizards, by contrast, provide perhaps the best examples of convergent evolution, for analogous species are identifiable between continents throughout the group. One-to-one matches are easily made on the basis of relative body size, behavior, and feeding biology, and show that niches similar in size and position relative to competitor niches within the community come to be occupied by ecospecies that possess a whole syndrome of adaptive features in common. Not only is this one-to-one correspondence evident at the species level, but because of its consistency at that level it extends to three-to-three or five-to-five correspondence at the community level, and also expands to community-level statistics such as relative densities and body-size ranges. Where exceptions do occur, they are compensated by niche expansion in a species of the depauperate community. The expansion may take the form of an increased intraspecific size range, as in the simultaneously active lizards in *Liolaemus nigroviridis*, which offset the absence of a small, allospecific counterpart to *Uta stansburiana;* or it may take the form of an increased habitat range, as in the occupancy of both woodland and scrub by *Sceloporus occidentalis*, the species corresponding to the Chilean *Liolaemus tenuis* in the woodland and to *L. fuscus* in the scrub.

We can suggest two possible reasons why the niche correspondence in lizards is so precise. First, the total number of lizard species examined is small, never exceeding five in the habitats studied. This means that a change in just one species in a community makes a good deal of difference to the remaining species, because the niches abut each other in some dimension, and each species is closely linked in ecological character to every other species. Second, the particular lizards studied may exploit resources that are largely unavailable to nonlizards—active anthropods on broad surfaces, caught by sit-and-wait techniques. The birds that forage on such surfaces—tree trunks and the ground—cannot rapidly search out the more cryptically concealed insects, and their faster metabolism may preclude their using sit-and-wait tactics to the extent lizards use them. Thus, the lizards exploit and perhaps monopolize a discrete resource that is obviously replicable in kind (though not necessarily in abundance or in systematic detail) on different continents. The resource will always be there, at least over an extensive season, which lizards, with their "on-off" metabolism, can track; other consumer groups are unlikely to affect the resource, and the outcome is strong convergence.

In birds, which have about the same α-diversity as ants and perhaps four times the lizards' species number, some species pairs and guilds show precise niche alignment and extensive convergence in overall ecology, morphology, and behavior. Others do not. The same two factors cited above—total numbers of species involved, and the degree of monopoly of the group over its resources—apparently control the degree of convergent evolution attained.

Grassland birds not only show one-to-one correspondence at the species level, but are ordered into communities that likewise exhibit precise cross-continental counterparts (Cody, 1968, 1973). But these communities comprise only three or four species; as soon as the species count exceeds about six, analogs become difficut to recognize throughout the community, and in twenty species communities about one-half the species cannot be assigned to one-to-one correspondences. In taller woodland and forest habitats, which at saturation support over thirty species in distinct niches, very few species show any intercontinental correspondence or any substantial degree of convergence.

Yet some sets of bird species show convergence with reliability. The food specialists, such as nectivores, raptors, trunk-and-branch searchers, and ground vegetarians, ordered usually in two- or three-species guilds, often show exact correspondence. Such resources as the foliage and the flying insects, to which few if any other consumers besides birds have access, are also exploited by guilds of bird insectivores. Here, too, there is excellent correspondence, even though such guilds contain four to six species ("warbler-types" and fly-catchers). Correspondence at this level is in marked contrast to that of the bird species that forage on the ground for insects or seeds, where there is at least potential competition from whole range of other consumers, including ants and mammals.

The mammal species of Chilean and Californian scrub are perhaps more equivocal than birds with respect to convergent evolution, despite the small size of their communities. The larger numbers of California species span a much greater size range, with additions at both extremes but particularly at the smaller, compared to Chile. The introduced *Rattus* in Chile helps to fill out the size range by occupying a conspicuous size gap in the endemic fauna; as yet, a similar gap in the Californian size range has not been occupied. There are some similarities between species "analogs" in external morphology and corresponding lifestyle, but no really convincing one-to-one correspondence. For instance, the larger rodents—*Neotoma* in California, *Abrocoma* and *Octodon lunatus* in Chile—are similar in coloration, size, and body proportions. They are all nocturnal, live in shrubs, and eat shrubby vegetation. A second *Octodon*, however, *O. degus*, has different food habits (it prefers herbs), a different schedule (it is diurnal), and correspondingly a different body coloration (uniform brown rather than countershaded grey). Likewise, there are on both continents "vole-types"—*Microtus* in California, *Akodon* in Chile—with torpedo-shaped bodies and short ears, feet, and tails. They are all associated with grass and litter. There are also on both continents "mouse-types"—*Peromyscus* in California, *Phyllotis* and *Oryzomys* in Chile—with large ears and long tails. These are associated with shrub-living and more arboreal, less fossorial, habits. But the details of their diets do not correspond well, for some Chilean species are more insectivorous (recall here again that the ground stratum in Chile has more herbs and denser insect populations). And although the "ground squirrel" and "vole" analogs on the two continents are the only small mammals showing an appreciable amount of diurnal activity, the species on one continent still differ somewhat from their closest intercontinental counterparts in schedule details.

The source of a good deal of the nonalignment in small mammals may be the absence in Chile of the seed-eating, cheek-pouched heteromyid rodents of North America. In North Africa, gerbils and dipodids have converged impressively with the hetromyids in both morphology and ecology, but in Chile there are no evident matches. More serious, for the South American rodents, than the absence of a heteromyid morphology may be their inability to reduce water loss in semiarid regions and their inability to utilize metabolic water. The Argentinian genus *Eligmodontia* perhaps approaches such adaptations (M. Mares, unpubl.), but it is biogeographically isolated from the Chilean scrub by the Andes.

Another source of mismatching in small mammals might be a greater independence among various niche dimensions on which natural selection is acting. For example, predation is a major selective force in small mammals, and the niches in "escape space" (the options available to small mammals to reduce predation) might be occupied by taxa relatively independently of the occupation of food-resource niches by these same taxa. Thus, a species pair might occupy on one continent portions X of escape space and Y of food space, but portions X and Y' on another continent, whereas a second pair might occupy escape and food niches X' and Y', respectively, on the first continent and X' and Y' on the second.

We note again that in small mammals it is the specialist species that provide the most convincing parallels. An example is the fossorial rodents *Spalacopus* in Chile and the analogous gopher *Thomomys* in California, species that belong to different families but have evolved very similar adaptations to cope with their peculiar subterranean living.

A group of consumers that we did not study in detail is the larger mammal herbivores. Mule deer (*Odocoileus,* family Cervidae) are the common browsers in Californian chaparral; their Chilean equivalents are guanacos (*Lama,* family Camelidae). The two are similar in size, proportions, and coloration, but guanacos now occur only in more open countryside and depend more on grazing than browsing. There are two cottontail rabbits (*Sylvilagus*) in the California scrub, one common species at each of the four sites studied; in Chile the introduced *Oryctolagus* is common at the primary site and present elsewhere. The native jackrabbits (*Lepus*) occur uncommonly at the Californian sites, and introduced *Lepus* from Europe is similarly rare at the Chilean sites.

The common rabbit predator in California is the coyote, *Canis latrans;* in

Chile the fox, *Dusicyon,* is superficially very similar to, and is as common as, the Californian coyote, and is likewise partial to rabbits. There are also mustelid and felid predators in both countries, but they show extensive taxonomic similarities: related genera, such as weasels (*Mustela* and *Galictis*); congeneric species, such as hog-nosed skunks (*Conepatus*); and even conspecific populations, of the mountain lion (*Felis concolor*).

Comparisons among the avian predators bring to light similarly close taxonomic affiliations, and their value in testing the convergent evolution hypothesis is accordingly low. The mediterranean regions of both California and Chile support conspecific vultures (*Cathartes aura*), owls (*Bubo virginianus, Tyto alba, Speotyto cunicularia, Asio flammeus*), falcons (*Falco peregrinus, F. sparverius*), and kites (*Elanus leucurus*), as well as congeneric hawks (*Buteo jamaicensis, B. polysoma, Accipiter cooperi, A. bicolor*) and closely related genera of condors (*Vultur* and *Gymnogyps*). One-to-one correspondence, then, becomes a phenomenon maintained by closely related taxa, rather than one generated out of unrelated forms.

To complete this discussion of predators, mention must be made of snakes. In Californian chaparral there are roughly sixteen species of snakes, with sizes ranging from 0.2 to 2.0 m and diets that include frogs, birds' eggs, insects, small mammals, and other snakes. In California's other mediterranean habitats, total species number is more than doubled, to about thirty-five. In the whole of mediterranean Chile, and constituting the most glaring example of nonconvergent evolution we can find, there are only two snake species! Granted, they are both food generalists, and the two are both size- and color-polymorphic. And since one is broad-headed and the other narrow-headed, and one large and the other of moderate size, the two might accomplish almost as much predation between them as their thirty-five counterparts in California. That just two snakes occur in central Chile presumably has a good deal to do with the effectiveness of the Atacama Desert and the Andean cordillera as snake (or, far less likely, snake-food) barriers; three snake species, for example, reach their southern limit near the Chile-Peru border. But the question of why an adaptive radiation into many species, like that in the genus *Liolaemus,* has not occurred in these Chilean snakes needs an answer, especially since the two snake species appear to have diversified ecologically, without evolving new species, and since snakes should be dependent on similar climatic or elevational barriers as the lizards and show in general a similar lack of mobility. So long as questions like this remain unanswered, we can entertain no complacency concerning the hypothesis of convergent evolution.

Density Compensation within Taxa

Occasionally in some consumer groups, routinely in others, we find imperfect one-to-one species correspondence. Furthermore, even when all of the species can be assigned to one-to-one correspondences, we often find that the ecological counterparts fail to correspond *comprehensively* across the entire range of adaptations attained by these species, even though important phenotypic traits do indeed match. The net effect is that resources are allocated among species sets on different continents in slightly or even very different ways.

At equilibrium, the density of a consumer species should reflect two factors, the density or abundance of its food resources and the density of competing species with overlapping resource-utilization curves. Very often, density differences between intercontinental species counterparts reflect primarily the differences in food abundance. For example, ants are much more common in California than in Chile, which allows a higher overall consumption of ants by Californian lizards, and accordingly a higher relative abundance of lizards dependent on the ant food supply. The Chilean ant-eating flicker, *Colaptes pitiusm,* is likewise much scarcer than its congeneric Californian equivalent, *C. cafer.*

Many California shrub species depend on hummingbirds for their pollination, and in return provide the hummingbirds (chiefly Anna's hummingbird, *Calypte anna*) with nectar, enclosed in small, trumpet-shaped corollas. The abundance of such shrubs in California accounts for the ubiquity of Californian hummingbirds. In Chile, only such scarce but conspicuous shrubs as *Lobelia* and *Sophora* are so adapted, and these are used by the giant hummingbird, *Patagonas gigas,* which is considerably more local in its occurrence than is its northern counterpart.

More direct evidence of the relation between consumer density and the abundance of food supply is offered by an irrigated field in Chile, where the three passerine bird species present have achieved a combined density 2-1/4 times that of birds in strictly comparable but nonirrigated Chilean and Californian fields carrying the same number of species.

As we have seen, density compensation for low α-diversity can take the form of extended habitat occupancy by species living along those parts of the habitat gradient that support low-α communities. We have also observed the special form of density compensation exhibited by the Chilean lizard *Liolaemus nigroviridis,* in which individuals of small size are simultaneously active with older, larger animals in compensation for the absence of a small allospecific counterpart of the second California species. But the principal form of density compensation remains the increase or decrease of competitor species in the presence of more or fewer consumers of the shared resources, in terms either of species numbers or of densities.

Density compensation of this sort can be illustrated by analogous sets of bird species. On each continent six bird species are the main insectivores above ground level. When each species is matched with its closest cross-continental counterpart, the average difference in density (in bird pairs per acre) is a substantial 62 percent. The difference can be attributed to imprecise niche alignment, imperfect correspondence in niche size and position between analogs, and differences in the insect resources available on the two continents. When we order the six species on each continent into three pairs, the pairings representing closest ecological affinities, and sum the densities of the two species in each pair, the mean density differences between continents fall to 55 percent; repeating this process of lumping species adjacent in niche space, by ordering each group into two trios, we find that the difference between the two-group averages is just 10 percent.

Density compensation is of course more apparent when species numbers do not correspond. We have seen, for example, that there are nine Chilean bird species, and five Californian counterparts, that forage chiefly on the ground. Three species pairs provide reasonable one-to-one matches: the two conspecific quail *Lophortyx californicus* (introduced to Chile); the slow, searching finches *Pipilo fuscus* (California) and *Diuca diuca* (Chile); and the long-billed mimids *Toxostoma redivivum* (California) and *Mimus tenca* (Chile). In California the densities of the three "matched" species sum to 2.0 pr/ha, in Chile to 2.5 pr/ha; the average densities per species (D/S) are 0.97 and 0.83 pr/ha, respectively, a little greater than the overall D/S for all mediterranean habitats of 0.78 pr/ha. The six additional ground foragers in Chile constitute an additional 4.63 pr/ha, and average 0.77 D/S; in California the two additional ground foragers are much more abundant, with D/S of 1.20 pr/ha. Thus, the five Californian species total 5.33 pr/ha, whereas the nine Chilean species total far less than 9/5 this value—in fact, 7.10 pr/ha. Additional species, then, do inflate the total density in saturated communities, but add pairs per acre at a rate considerably less than the overall average of 0.78 pr/ha for each extra species. Density compensation exists here, but is imperfect (i.e., total D correlates with total S but is not maintained as a constant).

The two flycatcher species in the evergreen woodland of Chile, *Pyrope pyrope* and *Elaenia albiceps* (the same two species found in the chaparral), total between them 2.28 pr/ha. In California the three flycatcher species in the evergreen woodland—*Myiarchus cinerascens, Contopus sordidulus,* and *Empidonax difficilis*—total 2.00 pr/ha. Again, a lower D/S is found in the community with fewer species, but in this case the total density is slightly greater in the community with fewer species. This modest discrepancy reverts to a substantial imbalance in the other direction when the remaining aboveground bird insectivores are considered. There are nine such additional species in California (warblers, vireos, etc.) with an average D/S of 0.83 pr/ha; but in Chile the additional four species of insectivores also produce an average D/S of 0.80 pr/ha. Thus, we find no density compensation whatever in woodland insectivores, for additional species in the richer community increase total bird density in the same proportion as densities in the species-poor woodland.

We conclude that density compensation does occur, but neither in all guilds nor in all communities. Nor does density compensation act in an all-or-none fashion; at best, it can compensate only in part for aberrant matches in species numbers.

Density compensation within taxa might be obscured by niche overlap and by density-compensation effects between taxa. Such effects could account for cross-continental density differences between analogous taxa for which resource abundances can be shown to be equivalent and within which density compensation operates. For example, part of the increased density and diversity of Chilean ground-foraging birds over their Californian counterparts might be due to the reduced ant densities and diversities in Chile, which therefore make additional insect and seed resources available to the birds. Other potential effects between taxa might array, for example, hummingbirds against such insects as bees and flies, in competition for nectar sources; or carnivorous, searching lizards, such as the Chilean *Callopistes,* against the co-existing rodents *Akodon,* in competition for similar prey.

Summary of Density and Resource-Use Patterns

Because convergent evolution occurs at the species level, species can be matched in pairs across continents and assigned analogous ecological roles that are often characterized by convergently similar morphology, behavior, and general biology. Such convergence is most readily distinguished if (a) the numbers of species in analogous communities match. Community sizes are likely to match if the overall species pools in the two regions are both large, and if the communities on the two continents are similarly affected by such species-formation and distribution factors as geographic barriers, taxon mobility, and evolution rates. Convergence is most likely if (b) the number of species in each of the two analogous communities is small, and if each of the species in each of the two communities is a specialist consumer of a discrete set of resources that is itself general and replicable on the other continent. Further, we can observe a convergence (c) in resource use between intercontinental counterparts that show no obvious morphological convergence, but nonetheless exhibit analogous behavior, foraging activity, and diet, as may happen if there are severe phylogenetic constraints on morphological adaptation.

We may just as easily fail to observe convergent evolution, at the species level, if any of the conditions defined above is violated or if the resource bases of quasi-analogous communities do not match. But natural selection may still produce predictable and intuitively reasonable patterns even if species groups show no one-to-one correspondences. Low α-diversity may be compensated by an expanded use of the habitats adjoining such species-poor communities, and by density compensation or niche expansion within the communities. And even if species in analogous communities show poor numbers correspondence and little or no density compensation, we can still observe that trends among communities within one continent parallel the trends in analogous communities on another continent, for example in changing species numbers over a habitat gradient in parallel fashion.

CONCLUSIONS

In the two preceding sections we have discussed convergent evolution among consumers in diverse taxonomic groups at two levels: community diversity and species distribution, on the one hand, and species correspondence with ameliorating density effects, on the other. The results, abbreviated in Table 6-3, produce the following generalizations:

1. In every taxonomic group the most similar characteristic of intercontinental comparisons and the feature most closely regulated by convergent evolution is α-diversity, the number of species coexisting within small patches of uniform habitat.

2. Below the community level, similar species numbers may be organized in different ways: one-to-one species correspondence can occur where community sizes are small and more or less equal; unequal species correspondence is partially compensated by niche breadth and density compensation, such that total consumer densities are appreciably greater in species-rich communities only when consumer species number in the one community approaches double that in the other community.

3. Above the community level, total species number in a given taxonomic group (birds, say, across all habitats) is expected to coincide between continents only fortuitously, since such totals are determined by species-production and dispersal factors that are governed by topographic features combined with taxon-specific mobility, chance, and, perhaps, history.

4. Between, on the one hand, α-diversities under the strict control of natural selection via interspecific competition and, on the other hand, regional species totals S_T produced by factors largely outside the control of natural selection, we have two additional diversity measures, β-diversity and γ-diversity; β-diversity appears to be more strongly regulated by competition, γ-diversity to be more influenced by topographic and, especially, regional isolation. The β-diversities are always greater where the S_T on one continent is greater than that on the other, but intercontinental differences in β-diversity are much smaller than are the differences in S_T.

5. Intercontinental differences in γ- and β-diversity are always of opposite sign. This reciprocal relation is evidently associated with the very factors that produce and limit species totals S_T. We surmise that the supracommunity differences between California and Chile are attributable chiefly to the greater geographic accessibility of the California mediterranean region. This greater accessibility has ensured larger S_T, which in turn has resulted in increased β-diversity and maintained a low γ-diversity. Thus Chilean lizards, in exhibiting higher β-diversity and lower γ-diversity than the Californian lizards, furnish the exception that proves the generalization: high S_T is associated with high β-diversity and low γ-diversity.

TABLE 6-3 *Species Correspondence, Total Taxon Densities, and Diversity Measures in Matched Communities in Chile and California*

Taxon	Total Species S_T	One-to-one species correspondence	Total density	Point diversity	α-diversity	β-diversity	γ-diversity
Ants	3.2>	no	?>	1.3>	1.7>	2.7>	<2.5
Bees	±4>	no	?	?	1.4>	±2>	<±3
Lizards	<1.7	yes, most	±=	<1.2	=	<1.5	1.2>
Birds	=	yes, some	±=	1.3>	=	1.5>	<1.6
Mammals	2.3>	yes, few		<1.3	=	1.8>	<2.5
Relative strength of natural selection vs. chance, history, and/or topography:	weak	fairly strong	very strong	very strong	very strong	fairly weak	weak

Note: (> indicates California value greater than Chilean value; < indicates Chilean value greater than California value; figure encompassed by > or < indicates factor by which the larger value exceeds the smaller.)

Summary and Conclusions

H. A. Mooney
M. L. Cody

In this study, scientists of various disciplines assessed the similarities and differences of two widely separated pieces of geography—southern California and central Chile. The two areas were selected rather subjectively to be as similar as possible environmentally and yet to be so distant geographically as to preclude major genetic affinities between their biotas. The areas thus offer a reasonable basis for testing the hypothesis that evolution in similar environments leads to convergence.

Each scientist studied these areas from a particular disciplinary viewpoint; and each study was self-contained at some level, since each addressed particular questions of the degree of similarity between regions whether of landforms, climates, or organisms. At the same time the results of each study contributed to a more synthetic view of the degree of convergence between the Californian and Chilean ecosystems as well as offering a basis for determining the selective elements leading to convergence.

We established first that the biotas of the Californian and Chilean regions are indeed taxonomically distinct, and thus, presumably, share common ancestries to only a small degree.

Next, we attempted to quantify the similarity in the physical environments of the two regions, and encountered here our first conceptual and interpretative difficulties. At a gross level of assessment there are probably no other two regions in the world so widely separated geographically and yet so similar in their physical environments. The types and configurations of landforms in the two regions are virtually mirror images of each other across the equator; in general aspect their climates are identical, and they have been directly comparable in both direction and rate of change during the recent geological past. Onto the same physical settings is overlaid the same veneer of human influences, for Europeans introduced the same crops to both regions, employed the same patterns of land use, and brought in the same weed species at about the same point in historical time.

But on closer examination at greater resolution, the similarities diminish and an increasing dissimilarity becomes apparent. At these finer levels of resolution, the adage that no two places on earth are identical receives greater credence.

At this stage we asked, what precisely are the finer similarities and differences in the two environments, and how must they be interpreted in an evolutionary context? The regions have comparable seasonal gradients in climate—cool, wet winters, and summer droughts increasing in length and severity with decreasing latitude. It is that period of drought that imposes an evidently overriding influence on the adaptive character of the biotas. The forms and functions of the dominant plants in the two regions are arrayed in a remarkably comparable and parallel manner along parallel latitudinal and topographic gradients—presumably because drought duration is the chief regulator of primary production, to which many morphological, physiological, and phenological traits are directly related. The adaptive types that optimize carbon-gaining capacity over the season at any point on these drought gradients are severely limited in number and show strong convergence; these types are replaced by adaptive types of different character at successive positions along the drought gradient, and such shifts proceed in parallel fashion in California and Chile.

Have the differences in climate between the two regions, putatively of minor importance yet readily measurable, had a differential evolutionary impact on the producers? The Chilean thermal regimes, particularly in the inland areas, where sclerophyll scrub predominates, are more moderate than their Californian counterparts, since they are more strongly influenced by marine air masses. Coincident with this climatic difference is the finding that Chilean plants show less phenological seasonality than those of California, and the Chilean shrubs are all more sensitive to frost than their Californian counterparts. It would thus appear that climatic differences between regions are paralleled or tracked by differences in the adaptive responses of the plants.

A major difference between the two regions is the lack of Santa Ana-type winds in Chile. The devastating widespread fires periodically provoked by these winds in California are conspicuously lacking in Chile. It would appear that these climatic differences largely explain Chile's lack of a distinctive flora of fire-adapted annual plants. Yet there are alternative explanations, which include of course the possibility of nonconvergent evolution. Other possible explanations center around those differences in land use that do exist between California and Chile. For our experimental design there were, unfortunately, no landscapes either in California or in Chile from which influences by the activities of man could be unquestionably discounted. In Chile, to a far greater extent than in California, the vegetation has been modified continuously over hundreds of years by a mosaic of small, local disturbances. This pattern of small-scale but widespread usage of the Chilean vegetation opens up a multitude of possible interpretations of the differences between the patterns of vegetation on the two continents, including the lack of a fire-adapted Chilean flora. The Chilean vegetation is both more open and more diverse than the Californian, and the herbaceous component of the flora has greater temporal permanence in Chile than in California. All of these features could be ascribed to man's disruptions of the Chilean vegetation. Since these differences in plant-community composition exist despite the fact that the dominant plants are highly convergent in such features as leaf type, branching pattern, carbon allocation, and primary production per unit biomass, our interpretation of them as aspects of human disturbances becomes even more credible. The difference between continents is most apparent in the way convergently similar plant species are assembled into communities, and it is just such differences in community configuration that we would expect to result from perturbations by man. In this case it becomes impossible to distinguish between alternative explanations for community-level differences: perhaps the differences between continents are explained by nonconvergence, yet the observed differences bear out expectations that take into consideration the known differences of higher-resolution environmental measurement.

Despite the difficulties of interpreting the differences between California and Chile in plant diversity and local distributional patterns, we could document in detail the adaptive responses that particular plant species had made through evolutionary time to their similar physical environments. From these studies emerges our view that the general climatic environment places identical constraints not only on the evolution of an optimal carbon-gaining strategy but also on the apportionment of this carbon through seasonal time to various competing ends, such as vegetative growth vs. growth of reproductive parts. Such similarities set the stage for convergence at the consumer level, for although gross differences in vegetation structure may exist, the details of individual plant structures are closely coincident.

From our studies of the consumer organisms of mediterranean Chile and California, we conclude that there is abundant evidence for convergent evolution among the vertebrates. This convergence is documented at a variety of levels, from parallels in subspecific adaptations to one-to-one correspondence between analogous species to convergence in the way communities are organized to exploit similar resource bases by similarly interacting consumers.

In all three groups of vertebrates—mammals, birds, and lizards—convergence was most apparent at the α-diversity or species per habitat level. Species correspond most exactly in lizards, where the absolute numbers are lowest (three to five species per site); only one of three pairs of sites does not match exactly, but there four Chilean species provide ecological matches to five Californian species since the younger individuals of the Chilean *Liolaemus nigroviridis* correspond to the Californian *Uta* and the larger individuals of the same species are ecological matches to the Californian *Sceloporus graciosus*. Bird species numbers match nearly as well in the lower habitats (e.g. nineteen Chilean vs. seventeen Californian species in chaparral), but maximal species numbers, while closer for common taxa alone, are invariably higher in California than in Chile. Trends in α-diversity among sites are parallel on the two continents, with most lizards found in montane forest, most mammals in coastal scrub, and with bird species numbers increasing monotonically from low coastal scrub through chaparral to woodland habitats.

Patterns in the use of habitat within each site shows less convergence in all three vertebrate groups, but similarities are still very striking in the lizards. The slightly richer Chilean matorral bird fauna was characterized by narrower within-site habitat tolerances, and the considerably less diverse Chilean mammal fauna by much broader species distributions within habitats.

While virtually all of the lizard species could be paired in 1-1 correspondences across continents to ecological analogs, only the flycatchers and the foliage, trunk, and aerial insectivores showed similarly exact convergence among the birds. While some mammal species in ecologically isolated and morphologically demanding niches (e.g., substerranean burrowers) showed very precise morphological and ecological correspondence between Chile and California, the majority of the mammalian fauna could not be assembled into 1-1 species matches; differences in both diet and diurnal activity patterns prevented the identification of close analogies.

Throughout the vertebrate faunas of both continents, it is among the insectivorous species rather than among frugivorous, granivorous, or herbivorous species that we can identify the closest cross-continental convergences. The consumption of more insects by the Chilean mammals (which include no cheek-pouched species) and more seeds by a more dense and diverse group of granivorous birds in Chile are examples of cross-continental differences that reflect differences in the resource bases available to the consumers. Chile has a larger herbaceous component to its vegetation than does California, and this produces more seeds and more insects at ground level as well as more foliage for ground-level browsers. This discrepancy only marginally affects the largely insectivorous lizards, alters resources for birds for just the ground-level foragers, but has a more marked effect on omnivorous and ground-level mammals.

In some case where species fail to correspond 1-1 across continents, we can often associate species into groups or guilds of ecologically related consumers with adjacent or overlapping resource utilities; then it is often possible to point out cross-continental covergence between these guilds of species, in, e.g., total densities of individuals, range of resource utilization, even though species numbers in the guilds do not correspond. Along with such 1-2, 2-3, or 3-5 correspondence between species sets in guilds, we observe that a guild of fewer species can compensate for low species count and may be ecologically analogous to a larger guild by density compensation, by virtue of higher densities per species in the smaller guild. We observe that at higher trophic levels, where consumer biomass and the sizes of guilds of coexisting species are smaller, convergent evolution is prominent; raptorial birds and predatory, larger, mammals are two groups that provide striking analogies, but their utility in support of the convergence hypothesis is diminished because of the greater phylogenetic similarity across continents at higher trophic levels of wider ranging animals.

In contrast to vertebrates, we can find little evidence of convergent evolution in nonvertebrate consumers, although our results apply only to select insect groups. To help explain this lack of correspondence in insects, we refer to the obvious trend of poorer and poorer convergence at both species and community levels as the numbers of coexisting species in taxa or trophic level, increase. In fact insect diversity is often several orders of magnitude greater than the diversity of vertebrate consumer groups. With such large numbers of insect species there are far more possibilities for alternative niche juxtaposition. Further, there is a perhaps fundamentally different degree of coevolution between insect consumers and their food resources; such tightly coupled coevolutionary systems could necessitate close and specific ties between consumer and resource within continents, and diminish the likelihood of finding convergent similarities between continents. Vertebrates, with their more generalized tropic interactions and a more plastic behavioral repertoire at their disposal, illuminate by convergent evolution the general similarities (amount, distribution, quality) of their food resources between continents but mask any differences in food plants at chemical or micromorphological levels that might be critical to insects.

While we present evidence for convergent evolution in vertebrates at various levels of organization up to that of the community, there is no evidence that broader levels of organization—supracommunity species counts and distribution—exhibit any degree of convergence at all. Thus we find that such statistics as the total number of species in a large county-sized or country-sized area, or the rate of turnover of species (accumulation of new species relative to the dropping of old species) with changing habitat (β-diversity) often bear no resemblance whatever across continents. Whereas the structure and organization of a particular community, such as the lizards in a chaparral patch, and of any subsection of that community, are convergent and thus predictable across continents, it is much more difficult to predict what parts of the chaparral lizard community, if any, with what shifts in their ecologies, if any, will show up in the lizard community of, for instance, an adjacent succulent coastal scrub. And, while the number of lizard occupants for montane forest types may be accurately predicted from a recipe that employs information on vegetation and substrate structure, the same recipe working on both continents, there is no such easy formula for predicting the number of lizard species in a coastal mountain range, in the suburbs of a major city, or in a desert region of normally varied topography.

The lack of demonstrable convergence at levels above the community in our vertebrate consumers can have a number of explanations, most of which are highly speculative and between them illustrate some of the most challenging of unsolved ecological problems. We might suppose that the roles of history, of chance, of topographic relief relative to the dispersal abilities and habitat accommodation of the species in question must all be extremely important. In fact, all of those biotic and abiotic factors that affect speciation rates, species-dispersal rates, and distributional limits, together with the temporal variation in these factors over recent and not-so-recent geologic time, can be assigned important roles with some assurance. Exactly what these roles are we have still to discover, and only their contribution to biogeographic noise is certain.

Our concluding remarks might best be directed toward some obvious weaknesses in our studies, toward what is most urgent in what remains to be done. We consider that by far the most valuable additional insights can be gained by continuing these studies on another continent, most fittingly in the Mediterranean itself (e.g., Sardinia, Turkey). Each additional continent that is thus drawn into the comparison contributes more than just an additional data point; for since it can be compared with each preceding study site, its empirical and theoretical value is immediately enhanced. Since there are strong affinities between South Africa and Australia both in landforms and soil types, affinities that have similarly influenced comparable floras, a comparison between these two remaining mediterranean sites should be made, but a comparison between either of them and any of the three already discussed might be less valuable, since the selective matrix differs so substantially.

In conclusion, we feel that the degree of convergence at the primary-producer level between the arrays of communities studied in southern California and central Chile is substantial. Chiefly responsible for the phenomenon, we argue, is the great environmental similarity of the two areas and the consequently strong direct linkage between plant morphological and functional types with climate. With the consumers the direct adaptive linkages between climate and morphological types are weaker and more difficult to demonstrate. Above the community level, in both plants and animals, convergences become less obvious and are increasingly obscured by chance and historical effects.

Appendix: Program Participants with Affiliation at the Time of the Project

Stanford University, California
Celia Chu
Karen Dement
William Dement
Susan Garrison
Sherry Gulmon[T]
Barbara Lilley
Harold Mooney[P]
David Parsons[T]
Sandra Pitelka

University of California, Berkeley
F. Lynn Carpenter
Robert Colwell[P]
Eduardo Fuentes[T]
William Glanz[T]
James Hunt[T]

University of California, Los Angeles
David Bradbury[T]
Martin Cody[P]
Noel Diaz
Norman Thrower[P]
Susan Woodward

University of California, Riverside
Homer Aschmann[P]
Conrad Bahre[T]
J. Marvin Dodge[T]
R. Anthony Van Curen[T]

San Diego State University
David Albright[T]
Michael Atkins[P]
Mitch Beauchamp
Roberta Bebout

Lucille Wesley Becking
Ruth Botten
Shawna Brizzolara
Tom Castonguay
Mary Duffy
Edna Ehleringer
Kathleen Fishbeck
Salley Gemeroy
Rachel I. Hays
Robert L. Hays[P]
Steve Hutchison
Barry Hynum[T]
Albert Johnson[P]
Barbara Johnson
Chris Johnson
Susan Johnson
Bill Jow
Jon Keeley
Sterling Keeley[T]
David Krause
William Lawrence[T]
Janet Lee
Philip McBarron
Jackie McClanahan
Patricia Miller
Philip Miller[P]
Russel T. Moore
Edward Ng[T]
Pierre Pincetl
Stephanie Pincetl
Dennis Poole[T]
Chris Rieden

Chris Robbins
Steve Roberts[T]
Eda Roberts
Jeff Rogers
Norman Slade
William K. Smith[T]
Marsha Pohl Sutherland
Ronald Tremper
Mary Ann Westfall
Ed Wosika
Universidad Catolica de Chile, Santiago
Sandra Araya
Maria-Ester Aljaro
Guacolda Avila
Albert Damm
Fernando Riveros de la Puente
Hiram Estay
Juan Giliberto
Louis Gonzales
Ernesto Hajek[P]
Adriana Hoffman
Alicia Hoffman
Jochen Kummerow[P]
Paule Le Boulengé
Emiliana Martinez
Juan Domingo Molina
Gloria Montenegro
Eugenio Sierra Ràfols
Carlos Roveraro

Mercedes Salgado
Andres Seguy
Universidad Catolica de Chile,
Valparaiso
Vilma Avendaño S.
Francisco Saiz[P]
Haroldo Toro
Enrique Vasquez M.
Universidad Austral de Chile,
Valdivia
Francesco di Castri[P]
Vitali di Castri[P]
Jaime Hurtubia[P]
University of California, Santa Cruz
Patricia Lincoln[T]
Alison Moldenke
Andrew Moldenke[P]
John Neff[T]
Harvard University, Massachusetts
Otto T. Solbrig
Duke University, North Carolina
Chicita F. Culberson[P]
William L. Culberson[P]
Kenneth Gabbard
University of Texas
Tom Mabry[P]
Dan Difeo[T]
University of Arizona
Valmore LaMarche[P]

P—denotes principal investigator, T—thesis researcher

Persons serving as project advisers
Daniel Axelrod
Harry Bailey
W. Frank Blair
Daniel Janzen
Orie Loucks
Anthony Orme
P. Pratt
Peter Raven
Jonathan Roughgarden

Jonathan Sauer
Persons assisting in taxonomic
identifications
Paul Arnaud
George Bohart
Richard Bohart
Howell Daly
Norman Downie
George Eickwort
Agustín Garaventa

Albert Grigarick
Jack Hall
Paul Hurd
Wallace La Berge
Hugh Leech
Mario Ricardi M.
Charles Michener
Luis Peña
Fresia Rojas

Jerome Rozen, Jr.
Richard Rust
Evert Schlinger
Roy Snelling
Robbin Thorp
P. H. Timberlake
J. Richard Vockeroth
Otto Zöllner

References

Adam, D. F. 1967. Late-Pleistocene and recent palynology in the central Sierra Nevada, California, pp. 275-302. *In* E. J. Cushing and H. E. Wright, Jr. (eds.), Quaternary paleoecology. Yale Univ. Press, New Haven, Conn.

Alfonso A., J. 1909. Los bosques i su legislación. Imprenta Moderna, Santiago.

Aljaro, M., G. Avila, A. Hoffmann, and J. Kummerow, 1972. The annual rhythm of cambial activity in two woody species of the Chilean "matorral." Amer. J. Bot. *59*:879-885.

Allard, H. A. 1942. Lack of available phosphorous preventing succession on small areas on Bull Run Mountain in Virginia. Ecology *23*:245-253.

Ando, M. 1970. Litter fall and decomposition in some evergreen coniferous forests. Jap. J. Ecol. *20*:170-181.

Arléry, R. 1970. The climate of France, Belgium, The Netherlands and Luxembourg, pp. 135-160. *In* C. C. Wallén (ed.), World survey of climatology, vol. 5: Climates of northern and western Europe. Elsevier Publishing Co., Amsterdam, London, New York.

Aschmann, H. 1973. Distribution and peculiarity of Mediterranean ecosystems, pp. 11-19. *In* F. di Castri and H. A. Mooney (eds.), Mediterranean type ecosystems: Origin and structure. Springer-Verlag, New York, Heidelberg, Berlin.

Axelrod, D. I. 1973. History of the mediterranean ecosystems in California, pp. 433-509.

Axelrod, D. I. 1973. History of the mediterranean ecosystem in California, pp. 225-277. *In* F. di Castri and H. A. Mooney (eds.), Mediterranean type ecosystems: origin and structure. Springer-Verlag, New York, Heidelberg, Berlin.

Axelrod, D. I. 1975. Evolution and biogeography of Madrean-Tethyan sclerophyll vegetation. Ann. Missouri Bot. Gard. *62*:280-334.

Axelrod, D. I., and H. P. Bailey. 1969. Paleotemperature analysis of Tertiary floras. Paleogeography, Paleoclimatol. Paleoecol. *6*:163-195.

Avila, G., M. Lajaro, S. Araya, G. Montenegro, and J. Kummerow. 1975. The seasonal cambium activity of Chilean and Californian shrubs. Amer. J. Bot. *62*:473-478.

Bahre, C. J. 1973. Land use in the Santa Laura area. Technical Report 74-19, Origin and Structure of Ecosystems Project, International Biological Program.

Bahre, C. J. 1974. Relationships between man and the wild vegetation of the Province of Coquimbo, Chile. Ph.D. thesis. Univ. California, Riverside. (University Microfilms)

Baumhoff, M. A. 1963. Ecological determinants of aboriginal California populations. Univ. California Publ. in Amer. Archaeology and Ethnology. *49*: 155-236.

Beadle, N. C. W. 1954. Soil phosphates and the delimitation of plant communities in eastern Australia. Ecology *35*:370-375.

Beadle, N. C. W. 1966. Soil phosphate and its role in molding segments of the Australian flora and vegetation, with special reference to xeromorphy and sclerophylly. Ecology *47*:992-1007.

Budyko, M. T. 1956. The heat balance of the earth's surface. Gidrometeorologicheskoe izdatel'stvo (Leningrad). Izv. Akad. Nauk U.S.S.R. Ser. G 3: 17. [Translation PB131692 U.S. Dep. Commerce, Office of Technical Services, Washington, D.C.]

Cain, S. A. 1944. Foundations of plant geography. Harper, London.

Carlisle, A., A. Brown, and E. White. 1966. Litter fall, leaf production and the effects of defoliation by *Tortrix viridana* in a small oak (*Quercus petraea*) woodland. J. Ecol. *54*:65-85.

Cahney, R. W. 1936. Plant distribution as a guide to age determination. J. Wash. Acad. Sci. *26*:313-324.

Chaney, R. W. 1947. Tertiary centers and migration routes. Ecol. Mon. 17: 139-148.

Cody, M. L. 1968. On the methods of resource division in grassland bird communities. Amer. Natur. *102*:107-147.

Cody, M. L. 1970. Chilean bird distribution. Ecology *51*:455-464.

Cody, M. L. 1973. Parallel evolution and bird niches, pp. 307-338. *In* F. di Castri and H. A. Mooney (eds.), Ecological studies, 7. Springer-Verlag, New York, Heidelberg, Berlin.

Cody, M. L. 1974. Competition and the structure of bird communities. Monogr. Pop. Biol., 7, Princeton Univ. Press, Princeton.

Cody, M. L. 1975. Towards a theory of continental species diversities: Birds distributions over Mediterranean habitat gradients, pp. 214-257. *In* M. L. Cody and J. M. Diamond (eds.), Ecology and evolution of communities. Harvard Univ. Press, Cambridge, Mass.

Cook, S. F. 1943. The conflict between the California Indian and white civilization: The Indian versus the Spanish mission, I. Ibero-Americana. *21*.

Dement, W., and H. A. Mooney. 1974. Seasonal variation in the production of tannins and cyanogenic glucosides in the chaparral shrub, *Heteromeles arbutifolia*. Oecologia *15*:65-76.

di Castri, F. and H. Mooney, 1973. Mediterranean-type ecosystems: Origin coast of North America, pp. 21-36. *In* F. di Castri and H. A. Mooney (eds.), Mediterranean type ecosystems: Origin and structure. Springer-Verlag, New York, Heidelberg, Berlin.

di Castri, F. and H. Mooney. 1973. Mediterranean-type ecosystems: Origin and structure. Springer-Verlag, New York, Heidelberg, Berlin.

Dietz, R. S., and J. C. Holden. 1970. Reconstruction of Pangaea: Breakup and dispersion of continents, Permian to present. J. Geophys. Res. 75:4939-4956.

Dodge, J. M. 1972. Forest fuel accumulation—a growing problem. Science. 177:139-142.

Dodge, J. M. 1975. Vegetational changes associated with land use and fire history in San Diego County. Ph.D. thesis, Univ. California, Riverside. (University Microfilms)

Donoso-Barros, R. 1966. Reptiles de Chile. Ediciones de la Universidad de Chile, Santiago, Chile.

Drew, L. G. (ed.). 1972. Tree-ring chronologies of western America. III. California and Nevada. Chronology Series 1. Lab. Tree-Ring Research, Univ. Arizona, Tucson.

Dunn, E. 1970. Seasonal patterns of carbon dioxide metabolism in evergreen sclerophylls in California and Chile. Ph.D. thesis. Univ. Calif., Los Angeles.

Du Rietz, G. E. 1931. Life forms of terrestrial flowering plants. Acta Phytogeogr. Suecica 3:1-95.

Duvigneaud, P., and S. Denaeyer-De Smet. 1971. Cycle des elements biogenes dans les ecosystemes forestiers d'Europe, pp. 527-542. In P. Duvigneaud (ed.), Productivity of forest ecosystems. UNESCO, Paris.

Encina, F. A. 1954. Resumen de la historia de Chile. Imprenta Editorial, Zig-Zag, Santiago.

Flint, R. F. 1971. Glacial and quaternary geology. Wiley, New York.

Frank, E. C., and R. Lee. 1966. Potential solar beam irradiation on slopes. Tables for 30° to 50° latitude. U.S. Dep. Agric. For. Serv. Res. Pap. Rocky Mountain Stn. RM-18.

Fuentes, E. R. 1974. Structural convergence of lizard communities in Chile and California. Ph.D. thesis. Univ. California, Berkeley.

Fuentes, E. R. 1976. Ecological convergence of lizard communities in Chile and California. Ecology 57:3-17.

Fuenzalida Ponce, H. 1967. Clima, pp. 99-199. In Geografía económica de Chile. Corporación de Fomento de la Producción, Santiago.

Fuenzalida Villegas, H. 1965. Biogeografía Chapter 7. In Geografía económica de Chile. Corporación de Fomento de la Producción, Santiago.

Gates. D. M. 1962. Energy exchange in the biosphere. Harper and Row, New York.

Gates, D. M. 1965. Energy, plants and ecology. Ecology 46:1-13.

Gentilli, J. 1971. World survey of climatology, vol. 13: Climates of Australia and New Zealand. Elsevier Publishing Co., Amsterdam, London, New York.

Glanz, W. 1976. Comparative ecology of rodent communities in California and Chile. Ph.D. thesis. Univ. California, Berkeley.

Góngora, M. 1960. Origen de los "inquilinos" de Chile central, Editorial Universitaria, Santiago.

González Rodríguez, R. 1966. Fauna económica de Chile continental. Chapter 7. In Geografía económica de Chile, primer appendice. Corporación de Fomento de la Producción, Santiago.

Good, R. 1956. Features of evolution in the flowering plants. Longmans, London.

Gordillo, C. E., and A. N. Lencinas. 1972. Sierras Pampeanas de Córdoba y San Luis, pp. 1-39. In A. F. Leanza (ed.), Geologia regional Argentina. Acad. Nae. de Ciencias, Córdoba.

Graenicher, S. 1930. Bee fauna and vegetation of Miami, Florida. Ann. Ent. Soc. Amer. 23:153-174.

Graham, A. 1973. History of the arborescent temperature element in the northern Latin American biota, pp. 301-314. In A. Graham (ed.), Vegetation and vegetational history of northern Latin America. Elsevier, Amsterdam.

Griffiths, J. F. (ed.). 1972. World survey of climatology, vol. 10: Climates of Africa. Elsevier Publishing Co., Amsterdam, London, New York.

Gulmon, S. L. 1977. A comparative study of the grasslands of California and Chile. Flora. In press.

Hamilton, E. L., and P. B. Rowe. 1949. Rainfall interception by chaparral in California. Calif. Dep. Nat. Res., Div. For.

Hanes, T. L. 1971. Succession after fire in the chaparral of Southern California. Ecolog. Monogr. 41:27-52.

Harrison, A. 1971. Temperature related effects on photosynthesis in Heteromeles arbutifolia M. Roem. Ph.D. thesis. Stanford Univ.

Harrison, A., E. Small, and H. A. Mooney. 1971. Drought relationships and distribution of two mediterranean climate Californian plant communities. Ecology 52:869-875.

Harvey, R., and H. A. Mooney. 1964. Extended dormancy of chaparral shrubs during severe drought. Madroño 17:161-165.

Hastings, J. R. 1964. Climatological data for Baja California. Univ. Ariz. Inst. Atmos. Phys., Tech. Rep. Meteorol. Clima. Arid Regions 14.

Hellmers, J., J. S. Horton, G. Juhren, and J. O'Keefe. 1955. Root systems of some chaparral plants in southern California. Ecology 36:667-678.

Hernández, S. 1970. Geografía de plantas y animales de Chile, Editorial Universitaria, Santiago.

Heusser, C. J. 1974. Vegetation and climate of the southern Chilean lake district during and since the last interglacial. Quat. Res. 4:290-315.

Hoffstetter, R. 1972. Relationships, origins and history of the leboid monkeys and caviomorph rodents: A modern reinterpretation. Evol. Biol. 6:323-347.

Holdridge, L. R. 1947. Determination of world plant formation from simple climatic data. Science 105:367-368.

Hünicken, M. 1966. Flora terciaria de los estratos del rio Turbio, Santa Cruz. Reta. Fac. Cienc. exact. fis. nat. Univ. Córdoba, ser. Cienc. Nat. 27:139-277.

Hunt, J. 1973. Comparative ecology of ant communities of California and Chile. Ph.D. thesis. Univ. California, Berkeley.

Jenkins, G. M., and G. D. Watts. 1968. Spectral analysis and its applications. Holden Day, San Francisco.

Jenkins, G. M. and G. D. Watts. 1974. Paleoclimatic inferences from long tree-ring records. Science *183*:1043-1048.

Jones, R. 1968. The leaf area of an Australian heathland with reference to seasonal changes and the contribution of individual species. Aust. J. Bot. *16*:579-588.

Keeley, S., and A. W. Johnson. 1976. A comparison of the patterns of herb and shrub growth in comparable sites in Chile and California. Amer. Midl. Nat., in press.

Keller, C. 1952. Introducción. *In* José Toribio Medina, Los aborigenes de Chile. Fondo Histórico y Bibliográfico José Toribio Medina, Santiago.

Kira, T., and T. Shidei. 1967. Primary production and turnover of organic matter in different forest ecosystems of the western Pacific. Jap. Jour. Ecol. *17*:70-87.

Kittredge, J. 1955. Litter and forest floor of the chaparral in parts of the San Dimas Experimental Forest, California. Hilgardia *23*:563-596.

Köppen, W., and R. Geiger, 1930. Handbuch der Klimatologie. Borntraeger, Berlin.

LaMarche, V. C., Jr. 1973. Holocene climatic variations inferred from treeline fluctuations in the White Mountains, California. Quat. Res. *3*:632-660.

LaMarche, V. C., Jr. 1974. Paleoclimatic inferences from long tree-ring Science *183*:1043-1048.

Leyton, L., E. R. C. Reynolds, and R. B. Thompson. 1967. Rainfall interception in forest and moorland, pp. 163-178. *In* W. E. Sopper and H. W. Lull (eds.), Forest hydrology. Pergamon Press, New York.

Linés Escardo, A. 1970. The climate of the Iberian Peninsula, pp. 195-240. *In* C. C. Wallén (ed.), World survey of climatology, vol. 5: Climates of northern and western Europe. Elsevier Publishing Co., Amsterdam, London, New York.

List, R. J. 1963. Smithsonian meteorological tables. Smithsonian Misc. Coll. 114.

Lossaint, P. 1973. Soil-vegetation relationships in Mediterranean ecosystems of southern France, pp. 199-210. *In* F. di Castri and H. A. Mooney (eds.), Mediterranean type ecosystems: Origin and structure. Springer-Verlag, New York, Heidelberg, Berlin.

Loveless, A. 1962. Further evidence to support a nutritional interpretation of sclerophylly. Ann. Bot. N. S. *26*:551-561.

MacArthur, R. H. 1970. Species packing and competitive equilibria for many species. Theor. Pop. Biol. *1*:1-11.

MacArthur, R. H. 1972. Geographical ecology. Harper & Row, New York.

MacArthur, R. H., J. MacArthur, D. MacArthur, and A. MacArthur. 1973. The effect of island area on population densities. Ecology *54*:657-658.

McBride, G. M. 1936. Chile: land and society. American Geographical Society, New York.

MacDonald, J. E. 1956. Variability of precipitation in an arid region: A survey of characteristics for Arizona. Univ. Ariz. Inst. Atmos. Phys. Tech. Rep. 1.

May, R. M. 1975. Patterns of species abundance and diversity. *In* M. L. Cody and J. M. Diamond (eds.), Ecology and evolution of communities. Harvard Univ. Press, Cambridge, Mass.

May, R. M. and R. H. MacArthur. 1972. Niche overlap as a function of environmental variability. Nat. Acad. Sci. Proc. *69*:1109–1113.

Medina, J. T. 1952 [1882]. Los aborigenes de Chile. Fondo Histórico y Bibliográfico José Toribio Medina, Santiago.

Mehringer, P. J., Jr. 1965. Late Pleistocene vegetation in the Mohave Desert of southern Nevada. J. Ariz. Acad. Sci. *3*:172–188.

Mehringer, P. J., Jr., and C. W. Ferguson. 1969. Fluvial occurrence of bristlecone pine (*Pinus aristata*) in a Mohave Desert mountain range. J. Ariz. Acad. Sci. *5*:284–292.

Menendez, C. A. 1969. De fossilen floren Südamerikas, pp. 519–561. *In* E. J. Fittkau, J. Ilies, H. Klinge, G. H. Schwabe, and H. Sioli (eds.), Biogeography and ecology in South America. Dr. W. Junk, The Hague.

Menendez, C. A. 1972. Paleofloras de la Patagonia, pp. 129–184. *In* M. J. Dimitri (ed.), La region de los bosques Andino-Patagonicos. Col. Cient. del INTA, Buenos Aires.

Mercer, J. H. 1972. Chilean glacial chronology 20,000 to 11,000 carbon-14 years ago. Some global comparisons. Science *176*:1118–1120.

Meserve, P., and W. Glanz. 1977. Rodent communities of arid regions of Chile. In prep.

Michener, C. D. 1954. The bees of Panama. Bull. Amer. Mus. Nat. Hist. *104*: 1–175.

Mikesell, M. 1960. Deforestation in northern Morocco. Science *132*:441–448.

Miller, P. C., and H. A. Mooney. 1974. The origin and structure of American arid-zone ecosystems. The producers: Interactions between environment, form, and function, pp. 201–209. *In* Proc. First International Congress of Ecology. The Hague, Netherlands.

Moldenke, A. R. 1971. Studies on the species diversity of California plant communities. Ph.D. thesis. Stanford Univ., Stanford, Calif.

Moldenke, A. R. 1975. Niche specialization and species diversity along an altitudinal transect in California. Oecologia *21*:219–242.

Moldenke, A. R. 1976. Evolutionary history and diversity of the bee faunas of Chile and Pacific North America. Wasmann J. Biol. *34*: in press.

Moldenke, A. R. 1976. Pollination ecology as an assay for ecosystem organization: Convergent evolution in Chile and California. MSS.

Moldenke, A. R., and J. L. Neff. 1974. The bees of California: A catalog with species reference to pollination and ecological research. Int. Biol. Prog. Origin and Structure of Ecosystems Tech. Rep. No. 74-1 – 74-6.

Moldenke, A. R., and H. Toro. The bees of Chile. In prep.

Monk, C. D. 1966. An ecological significance of evergreeness. Ecology *47*: 504–505.

Montané, M. J. 1968. Paleo-Indian remains from Laguna de Tagua-Tagua, central Chile. Science. *161*:1137-1138.

Mooney, H. A. 1972. The carbon balance of plants. Ann. Rev. Ecol. and Systematics. *3*:315-346.

Mooney, H. A. 1975. Plant physiological ecology—A synthetic view, pp. 138-150. *In* J. Vernberg (ed.), Physiological ecology. Intext Publ. New York.

Mooney, H. A. and B. Bartholomew. 1974. Carbon balance of two Californian *Aesculus* species. Bot. Gaz. *135*, 306-313.

Mooney, H. A. and C. Chu. 1974. Seasonal carbon allocation in *Heteromeles arbutifolia,* a California evergreen shrub. Oecologia *14*:295-306.

Mooney, H. A. and E. Dunn. 1970. Photosynthetic systems of mediterranean climate shrubs and trees of California and Chile. Amer. Nat. *104*:447-453.

Mooney, H., E. L. Dunn, F. Shropshire, and L. Song. 1970. Vegetation comparisons between the mediterranean climatic areas of California and Chile. Flora *159*:480-496.

Mooney, H. A., A. T. Harrison, and P. A. Morrow. 1975. Environmental limitations of photosynthesis on a California evergreen shrub. Oecologia *19*: 293-301.

Mooney, H. A., and R. I. Hays. 1973. Carbohydrate storage cycles in two California mediterranean-climate trees. Flora *162*:295-304.

Mooney, H. A. and J. Kummerow. 1971. The comparative water economy of representative evergreen sclerophyll and drought deciduous shrubs of Chile. Bot. Gaz. *132*:245-252.

Mooney, H. A., D. Parsons, and J. Kummerow. 1974. Plant development in Mediterranean climates, pp. 255-267. *In* H. Lieth (ed.), Phenology and seasonality modeling. Springer-Verlag, Berlin, Heidelberg, New York.

Mooney, H. A., J. Troughton, and J. Berry. 1974. Arid climates and photosynthetic systems. Carnegie Institution Year Book 73:793-805.

Morrow, P., and H. A. Mooney. 1974. Drought adaptations in two Californian evergreen sclerophylls. Oecologia *15*:205-222.

Muñoz Pizarro, C. 1966. Sinopsis de la flora Chilena. Universidad de Chile. Santiago.

Muñoz Pizarro, C., and E. Pisano Valdés. 1947. Estudio de la vegetación y flora de los Parques Nacionales de Fray Jorge y Talinay. Agric. Tec. 7:71-190.

Munz, P. A., and D. D. Keck. 1959. A California flora. University of California Press. Berkeley, Calif.

Ng, E. 1974. Soil moisture relations in chaparral. M. S. Thesis. San Diego State Univ., San Diego, Calif.

Nobs, M. A. 1963. Experimental studies on species relationships in *Ceanothus.* Carnegie Inst. Wash. Publ. 623.

Noy-Meir, I. 1973. Desert ecosystems: Environment and producers Ann. Rev. Ecol. Syst. *4*:25-51.

Oberdorfer, E. 1960. Pflanzensoziologische Studien in Chile. Flora et vegetation mundi, 2. J. Cramer, Weinheim.

Oficina Meteorologica de Chile. (1930-1970). Anuarios Meteorologicos de Chile. Santiago, Chile.

Osgood, W. H. 1943. The mammals of Chile. Fieldiana, Zoology 31:1-250.

Parsons, D. J. 1973. A comparison of vegetation structure in the mediterranean scrub communities of California and Chile. Ph.D. thesis. Stanford University, Stanford, Calif.

Parsons, D. J. 1976. Vegetation structure in the mediterranean climate scrub communities of California and Chile. J. Ecology. 64:435-447.

Parsons, D. J., and A. R. Moldenke. 1975. Convergence in vegetation structure along analogous climatic gradients in California and Chile. Ecology 56:950-957.

Paskoff, R. 1973. Geomorphological processes and characteristic landforms in the Mediterranean regions of the world, pp. 53-60. In F. di Castri and H. A. Mooney (eds.), Mediterranean type ecosystems: Origin and structure. Springer-Verlag, New York, Heidelberg, Berlin.

Patterson, B., and R. Pascual. 1972. The fossil mammal fauna of South America, pp. 247-309. In A. Keast, F. C. Erk, and B. Glass (eds.), Evolution, mammals and southern continents. Albany, State University New York.

Penman, H. L. 1948. Natural evaporation from open water, bare soil and grass. Proc. Roy. Soc. A 193:120-145.

Poole, D. K., and P. C. Miller. 1975. Water relations of selected species of chaparral and coastal sage communities. Ecology 56:1118-1128.

Raunkier, C. 1934. The life forms of plants and statistical plant geography. Oxford, Clarendon Press.

Raven, P. H. 1963. Amphitropical relationships in the floras of North and South America. Quart, Rev. Biol. 38:151-177.

Raven, P. H. 1973. The evolution of mediterranean floras, pp. 213-224. In F. di Castri and H. A. Mooney (eds.), Mediterranean type ecosystems: Origin and structure, Springer-Verlag, New York, Heidelberg, Berlin.

Raven, P. H., and D. I. Axelrod, 1974. Angiosperm biogeography and past continental movements. Ann. Missouri Bot. Gard. 61:539-673.

Reiche, C. 1896-1911. Flora de Chile. Imprenta Cervantes, Santiago de Chile.

Rodin, L., and N. Basilevic. 1968. World distribution of plant biomass, pp. 45-52. In F. Eckardt (ed.), Functioning of terrestrial ecosystems at the primary production level. UNESCO, Paris.

Romero, E. 1973. Ph.D. dissertation. Museo La Plata. Argentina.

Roughgarden, J. 1972. The evolution of niche width. Amer. Natur. 106:683-718.

Rowe, P. B., and R. A. Colman. 1951. Disposition of rainfall in two mountain areas of California. U.S. Dep. Agric. For. Serv. Tech. Bull. 1018.

Schimper, A. F. W. 1898. Pflanzengeographie auf physiologischer grundlage. G. Fisher, Jena.

Schlegel, F. 1962. Hallazgo de un bosque de cipreses cordilleranos en la Provincia de Aconcagua, pp. 43-46. Bol. Univ. Chile 32.

Schneider, H. 1969. El clima del Norte Chico. Universidad de Chile, Facultad de Filosofía y Educacion, Departamento de Geografía, Santiago.

Schulze, B. R. 1972. South Africa, pp. 501-586. *In* J. F. Griffiths (ed.), World survey of climatology, vol. 10: Climates of Africa. Elsevier Publishing Co., Amsterdam, London, New York.

Sellers, W. D. 1965. Physical climatology. Univ. Chicago Press, Chicago, Ill.

Shachori, A. Y., and A. Michaeli. 1965. Water yields of forest, maquis and grass covers in semi-arid regions. A literature review, pp. 467-477. *In* F. D. Eckardt (ed.), Methodology of plant ecophysiology. UNESCO, Paris.

Shantz, H. L. 1947. The use of fire as a tool in the management of brush ranges of California. Calif. Div. Forestry.

Shipek, F. 1968. The autobiography of Delfina Cuero. Morongo Indian Reservation, Malki Museum Press.

Simpson, G. G. 1950. History of the fauna of Latin America. Am. Scient. *1950*:361-389.

Simpson, B., J. Neff, and A. R. Moldenke. 1977. Flowers and flower visitors *In* G. H. Orians and O. T. Solbrig (eds.), Convergence in warm desert ecosystems. Dowden Hutchinson and Ross, Inc., Stroudsburg, Pa.

Slade, N., J. Horton, and H. A. Mooney. 1975. Yearly variation in the phenology of California annuals. Amer. Midl. Nat. *94*:209-214.

Small, E. 1973. Xeromorphy in plants as a possible basis for migration between arid and nutritionally deficient environments. Bot. Not. *126*:534-539.

Smith, G. I., E. B. Leopold, and E. L. Davis. 1967. Pleistocene geology and palynology, Searles Valley, California. Guidebook for Friends of the Pleistocene, Pacific Coast Section.

Snelling, R. R., and J. H. Hunt. 1975. The ants of Chile. Revista Chilena de Entomologia *9*:63-129.

Solbrig, O. T. 1976. The origin and floristic affinities of the South American temperate desert and semidesert regions, pp. 7-49. *In* D. W. Goodall (ed.), Evolution of desert biota. Univ. Texas Press.

Specht, R. 1969. A comparison of the sclerophyll vegetation characteristic of mediterranean type climates in France, California, and southern Australia. Aust. J. Bot. *17*:293-308.

Specht, R. 1973. Structure and functional response of ecosystems in the mediterranean climate of Australia, pp. 113-120. *In* F. di Castri and H. A. Mooney (eds.), Mediterranean type ecosystems: Origin and structure. Springer-Verlag, New York, Berlin, Heidelberg.

Stebbins, G. L., and J. Major. 1965. Endemism and speciation in the California flora. Ecol. Monogr. *35*:1-35.

Stebbins, R. C. 1966. A field guide to the western reptiles and amphibians. Riverside Press, Cambridge, Mass.

Thorne, R. F. 1973. Floristic relationships between tropical Africa and tropical America: a comparative review. F. J. Meggers, E. D. Ayensu, and D. Duckworth (eds.), pp. 27–40. Smithsonian Inst. Press, Washington.

Thornthwaite, C. W. 1948. An approach toward a rational classification of climate. Geog. Rev. *38*:55-94.

Thrower, N. J. W., and D. E. Bradbury. 1973. The physiography of the Mediterranean lands with special emphasis on California and Chile, pp. 37-52. *In* F. di Castri and H. A. Mooney (eds.), Mediterranean type ecosystems: Origin and structure. Springer-Verlag, New York, Heidelberg, Berlin.

Thrower, N. J. W., D. E. Bradbury (eds.). 1977. Chile-California Mediterranean scrub atlas: A comparative analysis. Dowden, Hutchinson & Ross, Inc., Stroudsburg. Pa.

Tosi, J. A. 1960. Zonas de vida natural en el Perú. Bol. Tec. No. 5, Inst. Int. Cien. Agr. DEA, Lima, Perú.

Tsutsumi, T. 1971. Accumulation and circulation of nutrient elements in forest ecosystems, pp. 543-552. *In* P. Duvigneaud (ed.), Productivity of forest ecosystems. UNESCO, Paris.

U.S. Forest Service, California Region. 1962. Multiple use management guides for the National Forests of Southern California.

U.S. National Oceanic and Atmospheric Administration. 1970-1974. Local climatological data: San Diego, California. Washington, D.C. Loose-leaf publ.

Van der Hammen, T. 1973. Historia de la vegetación y el medio ambiente del notre sudamericano, pp. 119-134. *In* 1° Congr. Lat. de Botanica Memorias de Symposio. Sociedad Botanica de Mexico, Mexico City.

Van der Hammen, T., and E. Gonzales. 1960. Upper Pleistocene and Holocene vegetation and climate of the "Sabana de Bogota" (Columbia, South America). Leids Geol. Med. *25*:261-315.

van Loon, H. 1972. Cloudiness and precipitation in the Southern Hemisphere, pp. 101-111. *In* H. van Loon, J. J. Taljaard, T. Sasamori, J. London, D. V. Hoyt, K. Labitzki, and C. W. Newton (eds.), Meteorology of the Southern Hemisphere. Amer. Meteorol. Soc. Meteorol. Monogr. 13.

Vuilleumier, B. 1971. Pleistocene changes in the fauna and flora of South America. Science *173*:771-780.

Wagenknecht, R. 1966. Catalogue of Chilean bees. Unpubl.

Wallén, C. C., and G. P. Brichambaut. 1962. A study of agroclimatology in semi-arid and arid zones of the Near East. FAO, Rome.

Watt, W., P. Hoch, and S. Mills. 1974. Nectar resource use by *Colias* butterflies. Chemical and visual aspects. Oecologia *14*:353-374.

Wells, P. V. 1969. The relation between mode of reproduction and extent of speciation in woody genera of the California chaparral. Evolution *23*: 264-267.

Wells, P. V., and R. Berger. 1967. Late Pleistocene history of coniferous woodland in the Mohave Desert. Science *155*:1640-1647.

Wells, P. V., and C. D. Jorgensen. 1964. Pleistocene wood rat middens and climatic change in the Mohave Desert: A record of juniper woodlands. Science *143*:1171-1174.

Whittaker, R. 1962. Net production relations of shrubs in the Great Smoky Mountains. Ecology *43*:357-377.

Whittaker, R. 1966. Forest dimensions and production in the Great Smoky Mountains. Ecology *47*:103-121.

Whittaker, R., and G. Woodwell. 1969. Structure, production and diversity of the oak-pine forest at Brookhaven, New York. J. Ecol. *57*:155-174.

Winnie, W. W. 1965. Communal land tenure in Chile. Annals Assoc. Amer. Geographers. *55*:67-86.

Wolfe, J. A. 1971. Tertiary climatic fluctuations and methods of analysis of Tertiary floras. Paleogeography, Paleoclimatol., Paleocol. *9*:27-57.

Wolfe, J. A., and E. S. Barghoorn. 1960. Generic change in Tertiary floras in relation to age. Am. J. Sci. *258A*:388-399.

Zinke, P. J. 1967. Soil and water aspects of brush conversion (Appendix B), pp. 21-27. *In* Proc. Calif. State Advisory Board. Sacramento, Calif. (Unpubl. ms.)

Zinke, P. J. 1973. Analogies between the soil and vegetation types of Italy, Greece, and California, pp. 61-82. *In* F. di Castri and H. A. Mooney (eds.), Mediterranean type ecosystems: Origin and structure. Springer-Verlag, New York, Heidelberg, Berlin.

Taxonomic Index

Subject Index